D0758226

# SEDUCED BY LOGIC

# SEDUCED BY LOGIC

*Émilie Du Châtelet, Mary Somerville and the Newtonian Revolution*

ROBYN ARIANRHOD

OXFORD
UNIVERSITY PRESS

Oxford University Press is a department of the University of Oxford.
It furthers the University's objective of excellence in research, scholarship, and education
by publishing worldwide. Oxford is a registered trade mark of Oxford University Press
in the UK and in certain other countries.

Published in the United States of America by
Oxford University Press
198 Madison Avenue, New York, NY 10016, United States of America



First published in Australia in 2011 by University of Queensland Press.



Library of Congress Cataloging-in-Publication Data

Arianrhod, Robyn.
Seduced by logic : Émilie Du Châtelet, Mary Somerville,
and the Newtonian revolution / Robyn Arianrhod.
p. cm.
ISBN 978-0-19-993161-3
1. Du Châtelet, Gabrielle Émilie Le Tonnelier de Breteuil, marquise, 1706-1749.
2. Somerville, Mary, 1780-1872. 3. Women scientists—France—Biography. 4. Women scientists—Great Britain—Biography. 5. Scientists—France—Biography. 6. Scientists—Great Britain—Biography. 7. Women—France—Intellectual life—18th century.
8. Women—Great Britain—Intellectual life—19th century. I. Title.
Q141.A725 2012
510.92—dc23[B] 2012006294

1 3 5 7 9 8 6 4 2

Typeset in Janson Text

Printed in the United States of America
on acid-free paper

# CONTENTS

# INTRODUCTION

Two of my favourite women in history are the wonderfully outrageous Émilie du Châtelet and the charmingly subversive Mary Somerville. Against great odds, Émilie and Mary taught themselves mathematics, and they did it so well that they each became a world authority on Newtonian mathematical physics. When I started out studying higher mathematics, the very existence of these women was enough to encourage me to believe that I, too, could succeed. But this book took shape much more recently, when I discovered a new connection with my heroines.

I was in Paris, researching Émilie's life and work, and just as I had hoped, it was both moving and exciting to be able to build a deeper kinship with her – walking the streets she walked, reading her poignant, passionate letters, and best of all, reading her painstakingly handwritten manuscript of her French translation of Newton's monumental *Principia*. But this last experience led me to something entirely unexpected: a deeper appreciation of Newton himself, and of the importance of his work not only in science, but also in our cultural history. Of course, I already admired Newton immensely: in the *Principia* (the full title of which is *Philosophiae Naturalis Principia Mathematica* or *Mathematical Principles of Natural Philosophy*), he published his famous theory of gravity, and in so doing, he created the very discipline of theoretical physics. But modern students and researchers generally study modern textbooks and current research papers, not fusty original sources, so it was through reading Émilie's translation of the *Principia* that I first seriously engaged with this magnificent book.

In fact, I became so entranced by Newton's awesome scope and style that, in order to appreciate it more fully, I began reading the *Principia* in English, and for many weeks I couldn't put it down. Predictably, I wrestled with the archaic and sometimes anarchic parts of it, but the underlying conception was bold and modern. I was astonished at Newton's extraordinary attention to detail, and I marvelled that one man was able to tie together, in one profoundly satisfying whole, all the astronomical knowledge of his forbears, as well as his own remarkable discoveries. And suddenly, I *felt* what Émilie must have felt 260 years earlier when she began her translation, because I, too, was in thrall to Newton's brilliance. Before then, my appreciation had been relatively academic, based on the fact that so much had flowed from Newton's work, including Einstein's theory of general relativity (which is my own area of research); but as I began to read the *Principia* more closely, I watched with awe and delight as the first comprehensive scientific theory in history took shape before my eyes.

We take the idea of universal gravity for granted today, so it is easy to forget that once it was actually considered contentious. However, in using his theory to solve the ancient mystery of why the planets move through the sky, Newton had wrenched humanity once and for all from its age-old place at the centre of the universe. Moreover, he had done it not with the traditional tools of theology or philosophy, but with the seductive logic of mathematics. Newton's mathematics was so seductive, in fact, that most of his late seventeenth-century Continental peers were both dazzled and suspicious, because the theory of gravity had been built solely on physical observation and mathematical insight, with no recourse to the usual religious or metaphysical hypotheses about the nature of reality.

But Émilie and some of her early eighteenth-century Enlightenment colleagues – including her lover, Voltaire – realised the *Principia* had changed not only the way we see the world, but also the way we do science. Newton had created a *method* for constructing and then *testing* theories, so the *Principia* provided the first truly modern blueprint for theoretical science as both a predictive, quantitative discipline – Newton eschewed qualitative, unproven metaphysical speculations – and a secular discipline, separate from religion, although by no means inherently opposed to it. Since then, this style of mathematical physics has had such a phenomenal impact on the way we live and the way we see ourselves in the universe that Newton is probably the most important scientist of all time, and Émilie was one of the first Continental scholars to actively promote his radical new way of thinking. In the late 1740s, she also made what is still the authoritative French translation of Newton's 510-page masterpiece, and she added a detailed 'commentary' that summarised both Newton's work and that of his earliest Continental followers.

Almost a century later, the Scottish-born Mary Somerville became a world authority on nineteenth-century Newtonian mathematical physics. There is an interesting symmetry in the fact that, while Émilie helped bring Newton's ideas from Britain to France in the first place, Mary helped take back to Britain the extraordinary Continental development of Newton's work that had taken place since then. In particular, while Émilie had translated the *Principia* into French, Mary translated into English the work of Pierre Simon Laplace, Newton's famous French disciple.

Laplace's exhaustive five-volume *Mécanique Céleste* (*Celestial Mechanics*) summed up the then-current state of gravitational mathematics. It was widely acclaimed as the greatest intellectual achievement since the *Principia*, and Mary Somerville was considered to be one of the few people capable of bringing it to an English-speaking audience. Her book was not a literal translation but an expanded version of Laplace's first two volumes, so it is a stand-alone exposition of the mathematics needed to explain what she called, in the title of her book, the underlying 'mechanism of the heavens'.

*Mechanism of the Heavens* became a standard text at Cambridge University for the next hundred years. And recently, it introduced me to Laplace's great work, although I am not interested in Mary and Émilie solely because they make accessible two of history's greatest men of mathematics. Nor do I claim they were original mathematicians themselves, although they were good enough to have had the respect of the greatest mathematicians of their time, and to write technical books that helped educate the elite young men who would become the mathematicians of following generations. Rather, I am fascinated by the way Émilie and Mary bring mathematics and physics to life, through their personal letters and recollections, and through their eclectic scientific and philosophical writings. They worked at a time when few women had any mathematical knowledge at all, and they were at the heart of the most advanced scientific society of their age. Consequently, they prove to be captivating, on-the-spot guides to two fascinating periods in scientific (and political) history.

Their greatest successes were in the areas of scientific translation and popular science writing – indeed, they helped pioneer the genre of serious popular science. But they also managed a degree of active involvement in the practice of physics, in which they published original research papers. (In fact, Émilie's experiments on colour and heat introduced me to Newton's other masterpiece, *Opticks*, which describes his meticulous experiments on the nature of light and colour.)

In my retelling of Émilie's and Mary's stories, one of my themes is the importance of mathematics as an idea, and its relationship with other ideas, especially those of philosophy and physics. But during the Enlightenment, many writers were inspired by the power of reason as exemplified by Newton himself – and such adulation turned him into the legend that still endures today. From a scientific point of view, he is indeed legendary, for the reasons I have sketched here and which I will explore further throughout this book. However, the story that unfolded in his wake had many twists and turns, because the historical process of science is much more complex and convoluted than might be expected from the detached, rational, "lone genius" stereotype of popular scientific legend. In fact, some of Émilie's Continental colleagues who initially rejected the theory of gravity went on to play a crucial role in bringing Newton's theory to its ultimate perfection.

As I have implied, Mary's era benefited from a century of post-Newtonian research, so that many of the scientific problems left unsolved in the eighteenth century were now resolved. Comparing Mary's life and work with Émilie's is a way of charting and appreciating these changes (and also some of the contemporaneous political and cultural changes, including the French Revolution and its aftermath). But above all, I want to highlight my heroines' curiosity – that wonderfully human desire to know and understand our world – and their enthusiasm for gaining such knowledge through logical thought, an enthusiasm they maintained despite the challenges and distractions of everyday life. In so doing, I hope to allow Émilie and Mary to take modern lay readers on a journey into the mathematical, scientific and philosophical ideas that seduced them both – and which, much later on, seduced me.

## *Chapter 1*

# MADAME NEWTON DU CHÂTELET

Émilie's story begins more than three hundred years ago, contemporaneously with the birth of mathematical physics itself. She was born Gabrielle-Émilie Le Tonnelier de Breteuil, on 17 December 1706, just nineteen years after the first publication of Newton's magnificent *Principia*. She was the daughter of the chief of protocol at Louis XIV's palace at Versailles, and she would marry into the prestigious Du Châtelet family. Voltaire would later refer to her playfully as 'Madame Newton du Châtelet', but she was far from a stereotypically staid female mathematician: aristocratic, sparkling with diamonds, adorned with silk and down 'pompoms' or other trinkets, she was as scandalous in her sex life as she was extravagant in her manner of dressing (Voltaire sometimes called her 'Madame Newton-Pompom-du Châtelet'). She had a 'temperament of fire', as she put it, a temperament that enabled her to live life to the full: at her château at Cirey-sur-Blaise, she could dance and sing entire operas all night long, and at the royal court at Versailles, she was a notorious gambler at cards. She felt that in both gambling and love, risking high stakes was a way of feeling fully alive.

There is very little reliable historical information about Émilie's childhood, although there are many anecdotes of her prowess at riding, and of her early preference for books over traditional 'female' interests. Certainly the adult Émilie was a fine horsewoman, and it seems she combined both the discipline and rationality of her mother,

Gabrielle-Anne de Froulay, and the passion and recklessness of her father, Louis Nicolas Le Tonnelier de Breteuil. In his earlier days, the baron de Breteuil had had a reputation for scandalous affairs, but by the time he married Émilie's mother, who was his second wife, he was forty-nine years old and had 'settled down'. Gabrielle-Anne was twenty-seven, and in later life she seems to have had an occasional brittleness that suggested she felt life had somehow passed her by. It was a fate Émilie was determined not to share. But first she would need to deepen her eduction.

According to a brief, second-hand reminiscence by Voltaire, when Émilie was in her late teens she was already fluent in Latin and could 'recite by heart the most beautiful passages from Horace, Virgil and Lucretius'. She was also familiar with all the works of Cicero, and had shown an 'early and dominant taste for mathematics and metaphysics'. Although this level of erudition was way beyond the pious and relatively rudimentary convent-based education received by most girls at the time, Émilie's later writings on the inadequacy of girls' education are so passionate and personal they suggest her own education was seriously lacking. Apart from lessons from her governess, she seems to have acquired much of her knowledge through her own reading, whereas her brothers were sent to one of the most prestigious Jesuit secondary schools; they could then attend university if they wished, but such a thing was unthinkable for a girl.

Fortunately, the Breteuil family environment was unusually favourable for self-education, as the unwitting evidence of the marquise de Créqui reveals. She had spent time as a child with the Breteuils – she was Émilie's cousin – and as an adult she wrote a memoir that painted the tall, studious young Émilie as pedantic and clumsy. Most scholars dismiss this as a caricature motivated by jealousy of Émilie's fame, but it is telling that Créqui had been astonished at the intellectual interests of her aunt and uncle, whose library alone 'usurped' three rooms. She also recorded anecdotes that suggest Émilie's parents were rather unorthodox in the intellectual freedom they allowed their children: both parents allowed Émilie to argue with them and express opinions, and from the time they were about ten years old, the children had permission to browse freely through the library.

While her education was unusually broad for a girl, the mathematics Émilie learned in her parents' house was relatively elementary. If she wanted to fully understand Newton's laws of nature and how they fitted in with free will – a question that tantalised the teenaged Émilie – she would need to study higher mathematics. Her struggle to learn such a difficult subject as an adult would elicit Voltaire's unstinting admiration – but that would come later because, at eighteen, Émilie dutifully entered an arranged marriage with thirty-year-old Florent-Claude, marquis du Châtelet and count of Lomont.

Within a year of the marriage, she gave birth to the couple's daughter, Gabrielle-Pauline, followed seventeen months later by their son, Florent-Louis. Six years later, another son would be born, but he died in his second year, causing Émilie unexpected grief. I say 'unexpected' because almost everyone lost children in those days of poor sanitation and nutrition, and also because upper class babies were immediately handed over to wet nurses, and no attempt was made to 'bond' mother and child. Émilie told a friend she was more upset at the baby's death than she would have believed possible, and that the 'sentiments of nature must exist in us without us suspecting'.

The marquis du Châtelet was a kindly man, and for about seven years after her marriage, Émilie happily played the role of society wife. But sometime around 1732, she experienced a true intellectual epiphany. Four years later, she would describe this intellectual awakening by saying 'chance' had led her to meet some 'people of letters, who befriended me, and I saw with extreme amazement that they took notice of me. I then began to see myself as a thinking creature'.

At first, she only caught a glimpse of this new possibility, and she continued to allow her time to be wasted by superficial society life and its dissipation, 'which was all I had felt myself born for'. Fortunately, her ongoing friendship with these 'people who think' – including another mathematically inclined woman, Marie de Thil, who would remain her lifelong friend – led Émilie to the liberating realisation that it was not too late to begin cultivating her mind seriously. But it would be a difficult journey: 'I feel all the weight of the prejudice that universally excludes [women] from the sciences. It is one of the contradictions of this world that has always astonished me, that there are great countries whose destiny the law permits us to rule, and yet there is no place where we are taught to think.'

In her attempt to become one of the 'people who think', she began to read more deeply. Presumably she studied Descartes, Newton, and the great English philosopher of liberty, John Locke (a favourite of Voltaire and other radical thinkers at that time), because when she met Voltaire a year after her epiphany, he was immediately captivated by her mind as well as her other charms. He described her to a close friend, Pierre Cideville: 'She is beautiful, and she knows how to be a friend . . . I swear to you, she has a genius worthy of Horace and Newton.' In an early poem to his new lover, Voltaire wrote, 'Ah! What happiness to see you, to hear you . . . and what pleasures I taste in your arms! I am so fortunate to love the one I admire . . . you are the idol of my heart, you make all my happiness.'

When Émilie and Voltaire began their courtship in 1733, she was twenty-six and he was thirty-eight (the same age as her husband, with whom Voltaire would eventually become good friends, thanks to Émilie's encouragement and her efforts as diplomatic go-between). Apart from a few poems and notes, it seems none of their love letters to each other survives – rumours suggest a jealous rival burned them. But unsubstantiated rumours have always surrounded the glamorous Émilie. From the very beginning, Voltaire contributed to her mystique by referring to her as 'the divine Émilie', but their relationship itself was shocking enough to bring notoriety. It was not the fact that she was married: aristocratic marriages were financial and dynastic alliances, not romances, and relationships between spouses tended to focus on duty rather than 'love'; consequently, extramarital affairs were the norm. But such affairs were not expected to be *grandes passions*; rather, they were to be conducted according to strict social rules that gave the illusion of fidelity to one's spouse, or at least of discretion, especially on the part of the woman. Émilie had developed a deep trust and friendship with the husband chosen for her by her parents and, for his sake, she did not want to bring scandal to the Du Châtelet name. Nevertheless, despite both her own best intentions and the fashion of the times, Émilie loved with her whole being, and she and Voltaire would not be able to remain discreet for long.

In associating openly with Voltaire, however, Émilie would break *two* social rules: not only that of discretion, but also that of class. She was the daughter of a favourite of the former Sun King, and she was the wife of a respected nobleman; Voltaire, on the other hand, had no aristocratic title. He was born François-Marie Arouet, but by the time of his relationship with Émilie, everyone called him by his pen name, Voltaire. Class differences were so important in pre-revolutionary France that a few years earlier, in 1726, he had been attacked with impunity by the chevalier de Rohan-Chabot, an aristocrat who regarded the up-and-coming playwright as a bourgeois upstart. In the foyer of the Comédie Française theatre, the two men had begun a bout of public verbal duelling, but few could match Voltaire's quick and caustic wit; when Rohan later had four henchmen beat Voltaire in the street, the outraged playwright had no legal redress, so he tried to provoke his adversary into a duel. Fortunately, if unjustly, the chevalier chose instead to use his family connections to have Voltaire thrown into the Bastille prison as a public nuisance.

He was released on the condition that he leave Paris until things calmed down, so he spent his exile in England, turning adversity into creative challenge by learning English, and by meeting some of Newton's disciples and learning about Newtonian science. In the way of great writers, he would ultimately put his newfound ideas into a book, *Lettres Philosophiques* (*Philosophical Letters*). Meantime, he enjoyed his cross-Channel sojourn so much that he stayed for two and a half years. One of the many things he admired about England was its constitutional monarchy, which had been established in 1689 after the 'Glorious Revolution': no doubt he envied the English their newly independent judiciary, free of interference from the king or royally privileged thugs like Rohan. In his *Letters*, he would write about the English political system, and about the ideas of Locke and

Newton, in a way that highlighted the intolerance and backwardness of French society and science. At the time he fell in love with Émilie, this project was not quite finished – it would be completed, with her help and support, over the coming months.

Aside from her controversial relationship with Voltaire, Émilie would also attract attention because of her intellectual passion. She hated the fact that women were denied the opportunity to fulfil themselves in this regard: not only did they have little access to formal education, but also they were allowed no chance to shape society, and thereby to 'taste glory', as she put it. So she was not content to become a 'person who thinks' such as one might find in a salon: she wanted to become a true scholar and, in particular, a mathematician. Indeed, underneath the flamboyance her nature was deeply serious, and not long after she fell in love with Voltaire, she wrote to a friend, 'Despite the pompoms, I think seriously about the welfare of my friends. I deliver myself up to the world without liking it very much.' She went on to say that playing society's games meant whole days passed by unnoticed, because in such a world one does not really feel one is living. And, unlike most women of her class and era, she worked hard to replace the emptiness of high society with scholarship: 'I love study with more fervour than I loved the world,' she wrote several years later, from her hideaway at Cirey.

Her study of Newton, in particular, shaped her life, and he is the main reason she is remembered today – although she is worth remembering for herself, too: she is both more spirited and more tragic than such great literary heroines as Anna Karenina. But Émilie's true essence cannot be understood without an appreciation of her intellectual passions, and of the intellectual climate in which she began her study. It is important, then, to review the significance of Newton's achievement, and the reasons it was so controversial.

*Chapter 2*

# CREATING THE THEORY OF GRAVITY: THE NEWTONIAN CONTROVERSY

In claiming the planets orbit the sun because of gravity, Newton was by no means the first to suggest the sun rather than the earth is the centre of our solar system: for the past century and more, mavericks like Nicolas Copernicus, Johann Kepler and Galileo Galilei had found suggestive evidence to support the idea. Most notably, Kepler had observed, from his painstaking analyses of Tycho Brahe's records of the planets' actual positions in the night sky at various times of the year, that the planets appear to move in mathematical ellipses around the sun. By contrast, when viewed from the earth, the planets' paths through the sky are erratic and complicated loops, with the various planets sometimes moving forward with respect to the earth and other times moving backwards. Kepler's beautifully simple ellipses provided an accurate description of planetary motion from a different point of view from the traditional earth-centred one – but which point of view was the 'real' one? Until someone found evidence that the earth was not fixed in space but physically moved through the sky, Christian doctrine on the matter would prevail in mainstream Europe, and the earth would remain at the centre of the universe. (This teaching arose from a literal interpretation of a Biblical reference to the sun moving about the earth.)

Nevertheless, most seventeenth-century philosophers accepted the existence of Kepler's meticulously observed ellipses as proof that the planets – including, by induction, the earth – really did move about the sun. After all, scientists had long supposed the simpler of two theoretical possibilities is likely to be the more correct. However, whether they moved around the earth or the sun, still the mystery remained as to why the planets should move in the first place. Newton is legendary because his solution to the riddle was so profound – and, leaving aside the mathematical details, so simple – that it seems awesome even today.

Building on the conceptual foundations laid by Galileo and Descartes, Newton began with three precise, axiomatic laws of motion. But then he did something no-one else had done: he proceeded to use his axioms to systematically articulate and prove scores of mathematical theorems about how and why physical objects move. In particular, he proved that for any object to move in an elliptical path, its motion *must* be caused by an 'inverse-square' force. This is a force whose measured strength is inversely proportional to the square of the distance between the source of the force and the point where its effect is measured, which means that if you double your distance from the source, then the force you feel is only a quarter as strong as it was initially, because the square of 2 is $2 \times 2$; if you treble your distance, the force is only $1/9$ as strong as it was, and so on. Newton also proved that the (approximately) circular motion of planetary moons is governed by an inverse-square force, too. Conversely, he showed that an inverse-square force will produce *only* elliptical or circular orbits – no other closed orbits are possible under such a force.

The important thing is that these were *mathematical* results. But Newton also used *experimental* data: in effect (and modernizing my terminology), he calculated the moon's acceleration around the earth, and compared it with Galileo's measurements of the rate at which objects fall here on earth – and he found that the moon's circular acceleration was just what you would expect if the moon were actually 'falling' towards the earth. This suggested that *gravity* was the source of the inverse-square force required to keep the moon in its circular orbit. In other words, if the earth's gravity extended as far as the moon – and if it obeyed an inverse-square law rather than being constant, as Galileo had supposed (acceleration due to the earth's gravity is constant only for bodies falling short distances from the same location) – then the moon's orbital motion could be explained by assuming it was caused entirely by the earth's gravity. Then Newton generalised his lunar result: knowing that the moons of Jupiter and Saturn move in the same way as the earth's moon, he deduced that all the planets must have gravity – of the same inverse-square type as the earth's. Finally, he concluded that the sun has gravity, too, and that the planets move in Keplerian ellipses because of their mutual gravitational attraction to the sun. On the surface, it is an astonishingly simple argument, but in truth it was an intellectual *tour de force* that would change forever the way scientists think.

Newton claimed that gravity is universal, by which he meant that every material body in the universe has gravity (not just planets and stars). Technically speaking, a body 'has gravity' if it exerts a gravitational force on other material bodies; it was a claim that would prove dangerously controversial, as Émilie's story will show. Meanwhile, to illustrate the power of his theory, Newton showed that it explained much else besides planetary motion. For instance, he was the first to explain the long-observed but long-inexplicable 'precession of the equinoxes', a slow rotation of the earth's axis. He did this with another piece of stunningly simple logic, arguing the inverse-square law shows that the moon (and to a lesser extent, the sun) pulls more strongly on the side of the earth that is closest to it at any given time, thereby 'nudging' the axis so that the spinning earth slowly precesses like a top. In a related argument, he showed how the combination of the sun's and moon's gravitational pull on the earth's oceans explains the tides, which scientists had been trying to understand for two thousand years; roughly speaking (focusing here on the dominant contribution of the moon), he showed that, as the earth rotates and the moon moves alternately closer and further from the ocean surface at any given location, the moon's gravitational pull on the ocean is alternately stronger and weaker, and so the tides wax and wane.

Newton also argued that comets – long feared by the superstitious – were nothing more than celestial bodies moving under the influence of gravity. He showed that some comets actually stay in the solar system, travelling on such extremely elongated elliptical paths that they return to our region only once every century or so.

Émilie later noted that Newton had deduced, from his mathematical law of gravity, what turned out to be more accurate descriptions of the planetary orbits than Kepler had found by watching the sky itself. In fact, it would take nearly two centuries, and the invention of powerful modern telescopes, to produce astronomical observations that were more precise than Newton's theory, which – when applied to the motion of planets and satellites in the solar system – turns out to be accurate to within one part in ten million.

However, Newton's theory was revolutionary not only because of its phenomenal accuracy, but also because of its dramatic philosophical significance. What Newton had done was to unite two apparently different physical phenomena – planetary motion and falling motion – using nothing more than physical intuition and mathematical logic. This gave a completely new insight into the nature of the universe: it showed that a single principle – that gravity is universal and acts according to an inverse-square formula – could explain so many physical phenomena that the world suddenly seemed comprehensible. At least, this is how Émilie and Voltaire saw Newton's legacy, and how we see it today. But when he first published his *Principia* in 1687, many of his contemporaries were not yet ready to be seduced by his mathematical brilliance.

Firstly, although the late seventeenth century was in many ways a scientific age, it was also a time when 'divine revelation' still had more official and popular clout than science. Besides, despite Newton's persuasive argument, direct evidence that the earth actually moves through the sky would not be found until just after his death in 1727. This was the so-called 'aberration of starlight', first analysed by James Bradley, Newton's friend and disciple. (The 'aberration' is the discrepancy between the observed direction of a star and its true direction, a discrepancy that Bradley explained in terms of the earth's orbital motion. It is analogous to driving in heavy rain: if the rain is actually falling vertically, then as you drive forward it will appear to fall towards you, as though it is falling from some place ahead of you rather than from directly above.) For Newton's scientific colleagues, however, the most controversial aspect of the theory of gravity was not its heliocentric focus but its mathematical nature, and it was this fact that aroused the suspicion of many physicists – or 'natural philosophers', as they were then called.

It was not the use of mathematics to *describe* nature that was the problem – the seventeenth century had seen an exciting consolidation of the role of mathematics as a descriptive language of nature: Kepler's planetary ellipses; Galileo's law of falling motion (namely, that all bodies near the surface of the earth fall at virtually the same rate, neglecting air resistance); Hooke's discovery of the mathematical law relating force and extension in a stretched spring; Huygens's discovery that pendulums of the same length all swing at the same rate regardless of the size of the swing or the mass of the pendulum bob (assuming the swings are relatively small, as in pendulum clocks, which were effectively invented by Huygens); and so on.

Obtaining mathematical descriptions of experimentally observed phenomena was a huge leap forward in physics, but no-one before Newton had created such a comprehensive, predictive, mathematical *theory* about the *underlying principles* behind such phenomena. In the case of planetary motion, understanding why the planets actually moved had been such a mystery that René Descartes, the early-seventeenth-century theorist whose ideas still dominated natural philosophy, had merely adapted Aristotle's ancient idea that space was filled with an invisible material 'ether', which supposedly contained vast, swirling vortices that dragged the planets around the sun.

Kepler had been closer to the mark when he spoke of magnetic-like 'emanations' from the sun, but this, too, was just speculation: neither Descartes nor Kepler had provided a rigorous, quantitative set of arguments to justify their theoretical ideas. But many of his colleagues did not take readily to Newton's pared down, mathematical approach: it subverted the traditional metaphysical paradigm in which causal suppositions, including beliefs about God, were the starting point for any attempt to *explain* the physical world rather than simply *describe* it.

For example, Gottfried Leibniz was Newton's most famous adversary. He was a first-rate mathematician himself – he was the independent co-inventor of calculus, Newton being the other inventor (although both men drew on the work of their immediate predecessors, including Cavalieri, Fermat and Wallis). But Leibniz rejected Newton's theory as unphilosophical and unintelligible, because it did not explain how his new concept of universal gravity actually worked: it did not explain how the sun could manage to hold the planets in orbit without any apparent contact. Following Descartes, Leibniz defined 'force' in terms of one object pushing or pulling or otherwise impacting on another via direct physical contact. Consequently, on reading the *Principia*, he wrote to Newton, 'You have made the astonishing discovery that Kepler's ellipses result simply from the conception of attraction or gravitation . . . And yet I would incline to believe that all these [planetary motions] are caused or regulated by the motion of a fluid medium' – namely, ethereal vortices that push and pull the planets around in their orbits. By contrast, Newton's gravitational force acted remotely, 'at a distance', without the aid of any intervening matter; in fact, he said the space between the sun and the planets was *empty* of matter.

Leibniz was so perplexed by the action-at-a-distance concept that he wrote to Newton's disciple Samuel Clarke, 'But what does he mean when he will have the sun attract the globe of the earth through an empty space? Is it God himself that performs it? But this would be a miracle, if ever there was any.' Which was another way of saying Newton did not have a proper theory. However, Leibniz went on to reject the idea of an inter-planetary vacuum in what we regard today as most unscientific terms: 'The more matter there is, the more God has occasion to exercise his wisdom and power – which is one reason, among others, why I maintain there is no vacuum at all.'

Besides Leibniz, Christiaan Huygens was one of the most important of Newton's scientific contemporaries, and he, too, was amazed at Newton's mathematical analysis of Kepler's laws. He told Newton it had never occurred to him to extend 'the action of gravity to such great distances as those between the sun and the planets, or between the moon and the earth, [because] the vortices of Mr Descartes – which formerly seemed to me rather likely, and that I still had in mind – hindered me from doing so'. But after he had recovered from his shock at Newton's mathematical brilliance, he told Leibniz that a force of gravitational attraction was an 'absurd' idea, and he, too, rejected Newton's theory because attraction was inexplicable in terms of mechanical principles – that is, in terms of an intelligible mechanism like push-pull impacts, as opposed to a mysterious action-at-a-distance force acting across millions of kilometres of empty space.

Newton himself had been bothered by the question of how gravity acted, but he saw it as beyond the scope of physics at that time. He believed that all we can know for certain about nature is what we are able to observe physically at the time, together with what we can deduce from these observations through creative but testable analogies between mathematical structures and physical phenomena. In other words, in creating his theory of gravity, Newton did not assume any hypotheses about the fundamental mechanism of planetary motion, or about the fundamental nature of the mysterious earthly force long known as 'gravity' (from the Latin for 'heavy'); rather, he simply studied the observed facts of planetary and falling motion. He then made analogies between mathematical ellipses or circles and the observed physical *paths* of planets or moons, and between the mathematical inverse square law and the observed *accelerating motion* of planets, moons and falling objects. But the key thing is that these mathematical analogies were testable, because they made quantitative predictions that could be confirmed by experiment; for example, using Newton's laws, physicists have successfully sent their own satellites into predetermined circular orbits.

It may sound simple enough today, but Newton's approach was nothing less than a new blueprint that defined modern theoretical physics itself, a paradigm shift in which the underlying principles of nature were to be inferred from the observed *effects* of natural phenomena, rather than from unproven assumptions about their *causes* – assumptions like vortices, or speculations about God. It was this radical methodological mix of precision and secularism that made the *Principia* so appealing to Enlightenment thinkers like Émilie and Voltaire, who were trying to define the boundaries between faith and reason, and, therefore, between religion and science, and between religion and politics. But it would take a great deal of study – and many twists and turns in her personal fortunes and her philosophical outlook – before Émilie became a world authority on Newton's controversial theory.

## *Chapter 3*

# LEARNING MATHEMATICS AND FIGHTING FOR FREEDOM

Émilie's plan to become a mathematician would require all her courage and determination. Firstly, envious acquaintances like Madame du Deffand would try to cast her as a dry and ugly 'learned woman' or *femme savante*, despite the fact that she had such appeal and charisma that the handsome duc de Richelieu, one of the most sought-after men in Paris, was rumoured to have once been her lover, while the celebrated Voltaire adored her.

Of course, some of her female contemporaries admired her scholarship: Madame de Graffigny would later say, 'Our sex ought to erect altars to her!' But many were irritated by, or envious of, her liberated commitment to an intellectual life, because Émilie was very different from the glamorous women who ran many of Paris's legendary literary salons. It was acceptable, even admirable, for such women to know enough of languages and philosophy to be good conversationalists with the learned men who dominated salon gatherings, but it was expected that women be modest about their knowledge. By contrast, Émilie would become famous as a scholar in her own right, thus angering the likes of Madame du Deffand, a powerful salonnière who claimed Émilie's interest in science was all for show.

Like most people, Du Deffand completely misjudged the depth of Émilie's intellectual commitment. Voltaire defended her by pointing out that such critics too easily bandied about the caricatures of Molière and Despréaux, who wrote dazzling satires on hypocrisy and pretension, but who implied – in Molière's *Femmes Savantes*, published in 1672, and Despréaux's *Satire X* of 1692, for example – that if a woman had scholarly ambitions, she was a *femme savante* with pretensions but no real scholarship. To be fair, Molière's earlier satire, *Les Précieuses Ridicules*, was apparently based on Madeleine de Scudéry, although recent scholarship suggests De Scudéry's writings were more sophisticated than her contemporaries had allowed. At any rate, there were so few truly learned women that the stereotype was based not on reality but on literary imagination, and yet, the idea that scholarly women were either pretentious or ugly would linger for the next three centuries.

In addition to the *femme savante* taunts, Émilie had to endure gossip that suggested she was interested not so much in mathematics as in having affairs with the men who taught her. Such gossip was inevitable, because her first teacher was the darling of the salon circuit, Pierre-Louis Moreau de Maupertuis. She began her first formal lessons with him late in 1733, a year after his rather cautious emergence as France's leading Newtonian. He had waited until his position in the Parisian Academy of Sciences was secure before going public with his views on Newton's theory, because the Academy was dominated by men conditioned by Descartes's notion of physics, in which forces act only through the direct contact of matter on matter. Like Leibniz before them, Maupertuis's Cartesian colleagues rejected Newton's notion of vast interplanetary vacuums penetrated only by some 'magical', disembodied force of gravitational 'attraction'. Consequently, Maupertuis had chosen London rather than Paris as the place to publish his first paper on Newton's theory, because 'I do not at all wish to read this piece in our [Paris Academy of Sciences] meetings where there are people who are shocked simply by the word "attraction".'

Among the Academy's Cartesians were such influential men as Bernard le Bovier de Fontenelle, the Academy's permanent secretary (a post held for life), and his future successor, Jean-Jacques Dortous de Mairan, with whom Émilie would later become embroiled in a scientific dispute. But the Continental rejection of Newton's theory was widespread, and it was not only Leibniz (who was German) and the French Cartesians who were repelled by the idea of gravitational 'attraction': men of the calibre of Huygens, who was Dutch, and the Swiss Jean Bernouilli, also refused to accept it. Bernouilli did accept Newton's inverse-square law, but he tried to show it could be derived not only from a remote force of gravity that *attracts* bodies to the sun, but also from a Cartesian system of invisible vortices of ethereal matter that *pushes* objects around the sun through a series of direct impulsions.

By the spring of 1732, Maupertuis had achieved both his coveted paid position in the Academy and the publication of his paper in the prestigious London-based *Philosophical Transactions of the Royal Society*, and he now felt 'brash' enough, as he put it, to go public in Paris with his cautious form of Newtonianism – namely, his belief that 'while waiting until we know what attraction is, I think it is evident enough from the facts that we can make use of it'.

Nevertheless, when it came to publishing his views in France, he had thought it safest – and more likely to enhance his reputation in the wider Republic of Letters – to bypass the Academy and append his mathematical paper to a semi-popular book in which he compared, in everyday language, the astronomical theories of Descartes and Newton. Although he tactfully left it for the reader to judge which one better fitted the observed facts, there were clear hints of his belief in the superiority of the latter's. His book, whose title translates as *Discourse on the various shapes of the celestial bodies, with an exposition of the systems of Mssrs. Descartes and Newton*, was published at the end of 1732, and it had made him quite a celebrity in literary and philosophical circles.

It was Maupertuis's support for Newton that had attracted Voltaire, who had been captivated by the new English physics since his exile in England. He had recently asked for Maupertuis's help in clarifying his ideas on gravity, because he was writing about it in his *Philosophical Letters*. 'Who would have thought, fifty years ago, that the same power would cause both the movement of the stars and [the phenomenon of] weight?' he had written to Maupertuis in March of 1733. By September, he was in love with both Newton and Émilie, and it was possibly he who helped tee up her lessons with the man he called the 'grand apostle of Newton'. It is likely he joined her initially, as a fellow student. (Later he and Émilie would take more advanced lessons from the young prodigy Alexis-Claude Clairaut, although both Clairaut and Voltaire soon realised Émilie was by far the superior student.)

At thirty-five, Maupertuis was both ambitious and charming. When he agreed to tutor Émilie, he probably expected her to be a dilettante like his other female students: he had quite a following among society ladies. But her first known letter to him, written in January 1734, is both deferential and eager: 'I spent all yesterday evening working on your lessons. I would like to make myself worthy of them. I fear, I confess to you, losing the good opinion you have of me.'

Perhaps he still doubted her commitment, because a week or two later she wrote, 'I spent the evening with binomials and trinomials, [but] I am no longer able to study if you do not give me a task, and I have an extreme desire for one.' Over the next few months, she sent him a stream of notes, trying to arrange lessons, asking him to come to her house for a couple of hours, or offering to meet him outside the Academy of Sciences – women were allowed inside only for the twice-yearly public lectures – or outside Gradot's, one of the favourite cafés of the intellectual set. Coffee and chocolate were relatively new in Europe then – Voltaire was reputed to drink forty cups a day of coffee mixed with chocolate – and the new cafés were another institution that was effectively closed to women.

It was this kind of intensity – as expressed in this multitude of requests for rendezvous – that fuelled gossip among her peers, and jealousy from Voltaire. Until the late twentieth century, most historians, too, seemed unable to imagine a woman like Émilie could be seduced only by mathematics – after all, until then, few women had actually become mathematicians. But it is true that many of Émilie's letters to Maupertuis have a very flirtatious style – it was, after all, an era that revelled in the game of seduction. There is no evidence to prove whether or not they ever became lovers in those early months, before she and Voltaire had fully committed themselves to each other, but her letters certainly prove that all her life she would continue to hold a deep affection and respect for Maupertuis.

In late April 1734, Émilie wrote to Maupertuis: 'I hope I will render myself less unworthy of your lessons by telling you that it is not for myself that I want to become a mathematician, but because I am ashamed of making such mediocre progress under such a master as you.' It was, indeed, an era of flattery! (Voltaire was quite adept at it – as a mere bourgeois, he often needed to flatter important people to help advance his literary career.) Although this letter suggests Émilie was simply using flattery to extract more lessons from her mathematical 'master', she always did have genuine doubts about her ability, which is not surprising given her lack of formal education and the assumed intellectual inferiority of her gender. She would later write, 'If I were king . . . I would reform an abuse which cuts back, as it were, half of humanity. I would have women participate in all human rights, and above all those of the mind.'

Émilie continued her chatty letter to Maupertuis by saying she hoped to hear from him, and that Voltaire sends his best wishes, although 'he is anxious, and with reason, about the fate of his [*Philosophical*] *Letters*'. It was a casual comment but, in hindsight, it was portentous. Voltaire's little book of popular essays on English institutions and philosophy would turn out to be one of the most influential and quintessential Enlightenment texts of the eighteenth century, with its emphasis on religious and political tolerance, and its support for philosophical and scientific empiricism as the most objective means of finding truth. Voltaire had composed it largely in English, and had published it in England, under the title *Letters Concerning the English Nation*, in August of the previous year. Soon afterwards, he had told a friend he had decided to hold off publishing the longer French edition, 'fearing more the clergy of the court of France than the Anglican church'.

In fact, Voltaire's anti-royalist, pro-tolerance views had already been dangerous for him. In 1717–18, he had spent eleven months in the Bastille prison, accused of having written one of his notoriously sarcastic pamphlets attacking those he felt had abused their power: it was such an attack that led Rohan to have Voltaire beaten and thrown in the Bastille in 1726, but back in 1717, his alleged target had been the duc d'Orléans, who was then the regent of seven-year-old Louis XV himself. (Orléans certainly had a tendency to personal dissipation, but it was his support for the charismatic Scottish speculator, John Law, that would later make him widely unpopular: Law had had some innovative ideas – Orléans instituted his idea of a national bank – but he overreached himself at the expense of his investors, including the French government, in a series of Mississippi land deals reminiscent of the wild housing and stock-market speculations of 2008–9: in the early 1720s, when the bubble inevitably burst, France suffered runaway inflation and economic disaster.)

Several years after Voltaire was released from the Bastille, the king's censors had banned his epic poem on Henri IV, which he had begun writing in prison; in 1723, the poem was printed in Rouen and smuggled into Paris, adding to his rebellious glamour in salon circles. A poem supporting Henri IV was problematic for the current king because, in 1598, Henri had signed the edict of Nantes, giving liberty of worship to Protestant Huguenots, a liberty that had been lost when Louis XIV revoked the edict in 1685. Huguenots had emigrated en masse to Holland and England in the bloody persecution that followed the revocation – Voltaire admired the relative tolerance of both these countries. Under Orléans – and then under Louis XV and his advisor, Cardinal Fleury – the government was less intolerant than it had been under Louis XIV, but nevertheless the status quo remained and Catholicism ruled, with no religious diversity allowed.

Apart from Louis himself, Cardinal Fleury was the most powerful man in France, and he and Voltaire were about to become antagonists over *Philosophical Letters*. In her letter to Maupertuis, Émilie's comment about Voltaire being 'anxious about the fate of his *Letters*' referred to the fact that they had recently received news that a pirated French version was circulating, although it had been published in London, thus bypassing the need for French royal authorisation. A few days later, the unthinkable happened: a French edition of the *Letters* appeared, published in Rouen but unauthorised by either

Voltaire or, more importantly, the royal censors who determined what was allowed to be published in France. As Émilie and Voltaire braced themselves for disaster, she realised just how much she loved him. He was 'the only person who has ever been able to fill my heart and my mind'. For the infatuated Voltaire, she was already the person whose 'friendship I prefer to all the rest' – but her practical, emotional and intellectual support during the drama that was about to engulf him would earn his lifelong gratitude. Soon after the unauthorised edition of the *Letters* appeared in May 1734, an arrest warrant was issued, and Voltaire went into hiding. Émilie was distraught, and she raged to Maupertuis against the injustice shown by his country to such 'an amiable and extraordinary' man.

Although there are no extant love letters between Émilie and Voltaire, her letters to others during the drama surrounding *Philosophical Letters* give us a glimpse of her intense way of loving. In a letter to an old friend – the abbé de Sade, uncle of the future marquis de Sade, who would become notorious half a century later – she poured out her heart, saying it was impossible to describe her sorrow at the plight of 'my friend Voltaire, for whom you know my feelings . . . His company is the happiness of my life . . . I tell you again, I don't think I can bear to think of him in prison, [where] he might die of sadness if not of illness . . . Madame Richelieu is my sole consolation here . . .' The reference to Madame Richelieu indicates the fact that Émilie was staying with the newly married Richelieus, where she and Voltaire and other guests had gathered for the wedding celebrations. She had been delighted to find her old friend's new wife was a warm and charming woman – Émilie had even taught her some mathematics in those happy spring days before disaster struck.

'I have no hope,' she continued wretchedly. 'Monsieur Chauvelin [the King's authorised representative who issued the arrest warrant] is inflexible and I am inconsolable.' In a remark that encapsulates both the sexual ethos of the time and the way she differed from it in her intensity, she added that the usual 'coquetry and other consolations for the loss of a lover will never heal this'.

A week later, things had taken a turn for the worse: she had not heard from Voltaire, and she feared how he would suffer on hearing the latest news – namely, that the Parisian parliament had formally denounced his *Letters*. Despite this news, on 7 June she told Maupertuis that Voltaire seemed happier, so that she was beginning to be able to think about other things, like mathematics. But she could not make progress with only a textbook: 'I believe it is only with you that I am able to learn, with pleasure, what A – 4A is equal to. You sow flowers on a path where others leave only thorns; your imagination knows how to embellish the driest of subjects, without leaving out their truth and precision.' This letter shows not only that Émilie was attracted to the 'truth and precision' of mathematics, but also the elementary level of her study at that early stage of her mathematical journey: today, students learn 'what A – 4A is equal to' in their second year at secondary school, just as Émilie's brothers would have learned it in school. The earlier reference to 'binomials and trinomials' suggests a similar, junior high-school standard.

Three days later, however, the royal executioner ceremonially burned Voltaire's incendiary little book, and Émilie believed they would have preferred to burn Voltaire himself.

What was it about *Philosophical Letters* that so upset the authorities? A short answer is that it was contemporary, lively, accessible writing that touched on some deep and controversial ideas. Voltaire was first and foremost a writer, a literary man rather than a scholarly author. But he was passionate about the fate of his country, which had been economically ruined after the wars of Louis XIV and Orléans's faith in John Law, and which was still divided by Catholic theological quarrels between the Jesuits and the fundamentalist Jansenists. (The Jansenists followed St Augustine in denying human free will a role in religious salvation.) In such a climate, Voltaire saw Britain as 'an island of reason'; in 1730, he had asked in despair, 'Is it only in England that mortals dare to think?' Émilie, too, admired the country of Locke and Newton: she would later write of 'my esteem for the

English, and the taste that I have always had for the free way of thinking and expressing themselves of this philosophical people'.

To emphasise the importance of religious diversity, Voltaire began his book with four articles (or 'letters') on Quakers, and one each on Anglicans, Presbyterians and Arians. He pointed out that Newton was an Arian, which meant he did not believe in the Trinity: in Voltaire's words, Arians believe 'the Father is greater than the Son'. Newton personally accepted Christ was a divine mediator between humans and God, but he was not God; to suppose otherwise was idolatry in his eyes. Voltaire claimed Newton was drawn to Arianism because it seemed more logical than conventional Christianity, and he also pointed out that Locke was a Nonconformist Christian.

Voltaire agreed that, privately, many in Britain detested each other's religious beliefs 'with nearly as much cordiality as a Jansenist damns a Jesuit in France', but he concluded, 'If there were only one religion in England, there would be despotism to fear; if there were two, they would cut each other's throats; but there are thirty of them, and they live in peace and happiness.' Or *relative* peace and happiness: in the last letter on Quakers, Voltaire noted that although members of the minority religions were not actively persecuted, they did suffer discrimination – they were not allowed to be members of parliament or take up any public office.

This was a theme he developed in the following letter, on Anglicanism, a letter that reveals Voltaire as a master of irony. It begins, 'This is a country that allows sects. An Englishman, as a free man, can go to Heaven by the path that pleases him.' Having thus attacked the French religious establishment, he then knocked down the English one: 'However, although each is free here to serve God in his own fashion, their true, most profitable religion is Anglican. It is impossible to get employment without being a faithful Anglican.' By which he meant employment at Oxford or Cambridge or in government office, for which prospective candidates had to prove their Anglicanism, by order of the so-called Test Acts. (Newton had kept his Arianism secret, although he drew the line at becoming Anglican in order to further his career. In the end, he obtained a special royal dispensation that allowed him to take up a chair at Cambridge without taking Anglican orders.) All is relative, however, and at the time Voltaire was writing, post-revolutionary Britain was far more tolerant than France: the Toleration Act of 1689 had given Nonconformist Protestants the freedom to have their own churches and schools, and taking occasional Anglican communion was often sufficient to satisfy the Test Acts. (Catholics, however, were perceived as such a threat to the Crown that they could not so easily bypass the Acts.)

After religion, Voltaire moved on to politics (notably Britain's constitutional monarchy), commerce, philosophy, science and literature, but the most controversial parts of the book were the letters on Newton and Locke. It was as if the authorities could see the future and desperately wanted to stop it, because Locke and Newton were, indeed, two of the most important influences on the subsequent political and intellectual development of the West.

In his brief popular overview of Newton's theory of gravity, Voltaire did not hide his support for Newtonianism as Maupertuis had tried to do. On the contrary, he included a specific response to the objections against the theory of gravity made by two of the most influential Cartesians in the French Academy of Sciences – namely, Fontenelle and Joseph Saurin. He noted that their rejection of the idea of 'attraction', simply because they did not know what caused gravity in the first place, was misguided, because neither did they know what caused the ethereal 'impulsions' that were their preferred 'explanation' for planetary motion. Voltaire said this misunderstanding was surprising, because Newton himself had told his readers he was dealing only with the demonstrable, quantifiable *effects* of gravity, leaving to posterity the task of uncovering its underlying metaphysical nature.

It was an incisive critique, although Maupertuis had already used it, much more tactfully (without naming names), in his *Discourse* of 1732: it was important in Parisian social and scholarly circles to be polite, and in his role as secretary of the Academy, Fontenelle discouraged members from directly criticising each other. However, although Voltaire's support for Newton was shocking in its directness, it was in defence of Locke that he was at his confrontational best – or worst, according to the mores of the reader.

Like Newton, Locke was a supporter of constitutional government, but it was Newton's new physics, rather than his politics, that prompted Locke to seek out a meeting with him in 1689. Not that Locke had studied all the complex proofs in the *Principia* himself – his specialties at Oxford had been humanities and medicine, not mathematics – but he had checked with Huygens, who reassured him that Newton's mathematics was sound. After meeting Newton, Locke immediately developed a close friendship with him, and Newton wrote out for his new friend a simplified paper on the mathematics of elliptical planetary motion.

Both Locke and Newton were influenced by the empiricism of the late seventeenth century – by the importance of physical facts and experiments in building up a picture of physical reality. In this matter, Locke was particularly inspired by the ideas of his predecessors Francis Bacon and Pierre Gassendi, his contemporaries Robert Boyle and Christiaan Huygens, and, above all, the 'incomparable' Newton. Locke's philosophy concerned both the nature of thought and the nature of liberty and good government. In examining the way we think – in *An Essay Concerning Human Understanding* – Locke expressed himself in the style of Newton, in the sense that he focused on logical deductions based on the everyday experience of thinking, rather than speculating on the nature of the mind itself. In other words, his goal was to understand the nature of ideas, not the mechanism or mind that produces those ideas in the first place, just as Newton had focused on deciphering the empirical law of gravity, not its underlying mechanical or metaphysical nature. However, at a time when thinking was seen as a function of the soul, Locke's *Essay* was controversial: in refusing to speculate on the underlying nature of the mind, it kept philosophy and religion separate, just as Newton's *Principia* had done.

Locke's friendship with Newton was perhaps an impetus finally to consolidate and publish his longstanding ideas: in 1690, two and a half years after the *Principia* appeared, Locke published both his major books, *An Essay*, and *Two Treatises of Government*. He published the latter work anonymously, presumably because of its revolutionary 'social contract' theory. The original 'social contract' concept was an attempt to articulate the nature of government, in which people agree to accept the authority of the state in exchange for protection against lawlessness. Locke maintained that political leaders should be subject to the same laws as everyone else, and that no government should deny individuals their basic 'natural' rights. He was not the first to make such points, but his work was cogent and timely, providing support for the idea of constitutional government and, implicitly, for the recent 'Glorious Revolution' that instituted Britain's constitutional monarchy. His predecessor Thomas Hobbes had believed that only an all-powerful government could protect individuals from the depredations of an innately selfish populace, but Locke believed that granting absolute power to any leader was dangerous, because such a sovereign could become 'licentious with impunity', as he put it. Consequently, he did not accept the 'divine right' of kings to rule as absolute monarchs, and he argued that it may be necessary to overthrow any government that did not respect its citizens' natural rights, such as the right to the fruits of one's labour, and to fundamental liberty. He believed everyone was born equal in terms of their rights, and that people were basically reasonable and ethical, so his notion of individual rights emphasised reciprocity and responsibility.

Locke's political theory would later inspire the authors of the revolutionary French and American declarations of human rights, but these revolutions were still half a century away when Voltaire wrote *Philosophical Letters*. He was more concerned with Locke's *Essay*, particularly its comments on the relationship between faith and reason. Locke said these should be able to coexist, and that it is only the wrong-headed *opposition* of faith and reason that leads to 'those absurdities that fill almost all the religions that possess and divide mankind', because even knowledge revealed by God must conform to reason. Otherwise, he cautioned, it is too easy for people to mistake enthusiasm, superstition or individual whim for divine revelation. But this was not the main reason that Locke outraged many religious authorities: *An Essay* led him into the dangerous territory not only of religious freedom but also of the soul.

Recall that for most people at that time, mind and soul were inseparably linked, but that Locke, like Newton, was wary of drawing conclusions from untested hypotheses. He argued that no-one knew the exact relationship between body, mind and soul – for example, whether the soul's existence began prior to conception, at conception, or some time afterwards. What we *do* know is the empirical evidence of our own experience, from which Locke deduced that 'although the soul may be continually present', he himself had 'one of those dull souls' that did not 'contemplate ideas continually'. Therefore, he concluded, for philosophers 'to say that actual thinking is essential to the soul, and inseparable from it, is to beg what is in question, and not to prove it by reason'. In fact, he believed no-one knew exactly what the mechanism or location of thought was, and he suggested that perhaps even a material being without a soul can think, although he acknowledged we may never be able to know this for certain.

Hobbes had already begun a materialist analysis in which he argued sensations were produced solely by motions in the various organs of the body, but Locke's analysis of thought was more thorough, and less mechanical. He personally felt it was inconceivable that thinking could take place without the aid of a soul or some immaterial agent, but to deny this possibility without evidence meant limiting God's power; after all, he said, we do not understand the fundamental mechanism of gravity, but it exists.

In *Philosophical Letters*, Voltaire noted that according to some theologians, Locke's suggestion that a purely material being might be able to think was akin to saying the soul itself was material and mortal. Voltaire pointed out that Locke was talking about thought, not the soul. As for the latter's presumed immortality, Voltaire claimed it was impossible to know whether or not this was true because no-one knew the soul's fundamental nature. But then he added, in his typically provocative style, that such philosophical distinctions were not made by the religious; instead, 'we are ordered to believe as a matter of faith: the thing is decided, [and] it matters little to religion what the nature of the soul is, as long as it is virtuous'.

Today, philosophers are still wondering whether or not consciousness exists outside the brain; however, for a modern perspective on the intensity of this debate in the seventeenth and eighteenth centuries, recall current arguments against abortion and embryonic stem cell research that are based on beliefs about the soul, with reference to when an embryo becomes human. The debate itself is important, and I do not mean to imply that a scientific or a secular opinion on such matters is automatically better than a religious one. Rather, the issue here is not the opinions themselves, but the freedom to express those opinions, on both sides of the debate: secular as well as religious, religious as well as secular. This aspect of the idea of liberty is as important today as it was in Locke's and Voltaire's time. Three centuries ago, however, Europeans were still occasionally burning so-called heretics, and Voltaire continued his letter on Locke chillingly:

> The superstitious say it is necessary to burn, for the good of their souls, those who suspect we think only with our material bodies . . . The superstitious are to society what cowards are to the army: they create panic. They cry that Locke wants to overthrow religion, but he has said nothing about religion in this matter – it is purely a question of philosophy, completely independent of faith and revelation. But then, theologians too often begin by claiming *God* is outraged when someone disagrees with their *own* opinion.

When he heard that *Philosophical Letters* had been officially declared 'scandalous, contrary to religion, morality and the respect due to the Authorities', the unrepentant Voltaire told his close friend Charles-Augustin de Feriol, comte d'Argental, 'Truly, they cry so much about these wretched *Letters* that I wish I had said more.'

*

Both Émilie and her husband had been actively interceding on Voltaire's behalf, but in the end it was Madame de Richelieu who successfully presented to the authorities a formal apology from Voltaire, together with his statement that he had had no involvement in the French publication of *Philosophical Letters.* The spectres of imprisonment or exile that had tormented Émilie were vanquished at last, because although Voltaire was banished from Paris and the royal court, he was given permission to live in the château at Cirey, on the Du Châtelet estate in Champagne. Presumably the authorities believed the upstanding marquis would keep the errant Voltaire in line, although he was rarely at home because of his military duties: all aristocratic men were expected to serve their king in a military capacity, and from 1733 until he retired from active service in 1749, the marquis du Châtelet was away most of the time, serving in the various European territorial battles of the era.

The estate at Cirey produced much of the marquis's income – through rents from tenant farmers, and through the estate's timber, vineyards and forges – but the château itself was rather rundown, having been uninhabited for years. Nevertheless, Voltaire exuberantly set about planning to restore it. He intended to pay for much of the renovation himself, because he was already quite wealthy: over his lifetime, he made a substantial amount of money as one of the bestselling authors of his age, but the foundation of his fortune had been laid a decade before he met Émilie, in a scam in which he and mathematician Charles La Condamine formed a syndicate that bought up all the tickets in a national lottery. Voltaire had invested his winnings astutely, and he continued to build his wealth by paying close attention to his investments. (He did not appear to worry too much about the business ethics of some of the companies in which he invested – he did not mind selling goods at inflated prices in captive markets, for example; as far as he was concerned, capitalist commerce not only delivered great benefits to society, it also went hand-in-hand with political liberty. He had expounded upon this theme in *Philosophical Letters*).

Émilie dreamed of fleeing to Cirey with her lover, but it was only a brief fantasy: the idea of retreating for long periods to the country, with Voltaire rather than her husband, would have seemed preposterous, even scandalous, to most of her friends. Cirey's isolation alone was enough to make the idea seem absurd: wolves still roamed its forests, and its solitude seemed frightening to city folk. In early October, however, Émilie arrived at Cirey with 'two hundred boxes', as Voltaire exaggerated it, but she was not yet bold enough to move there permanently: she intended to stay for two months. Voltaire was ecstatic, taking great delight in working with her to turn the château into a home, and planning to invite friends to come and share it. In a letter to one such friend he wrote, 'Amidst all this disorder, Madame du Châtelet laughs and is charming. She arrived here bruised and shaken after the long journey [of about four days in a roughly sprung carriage], without having slept, but she is well. She asked me to send you her best.'

In late October, Émilie wrote Maupertuis, 'I am here in this profound solitude, to which I am accommodating myself well enough; my time is divided between the masons and Mr Locke.' As for mathematics, she repeated her complaint that a textbook was not enough and that she needed Maupertuis or Clairaut to make the subject live for her; but they were in Switzerland, studying with one of the greatest mathematicians of the era: Jean Bernouilli. She went on to say, 'They write me from Paris that there is a priest of the Christian doctrine who is undermining and reducing to dust the system of Mr Newton: he doesn't realise you could strike him down if you believed him worthy of your anger; personally, I don't believe he is worth the effort.'

By December – at the end of Émilie's two-month sojourn at Cirey – Voltaire was still not free to return to Paris with her. But she made the best of it, enjoying city life again after the solitude of Cirey. On Christmas Eve she wrote an intriguing note to Maupertuis: '[I see you so rarely even though we are both back in Paris.] I want to celebrate the birth of Héloise with you; see if you want to come and drink her health this evening, with Clairaut and me. I will wait for you between 8 and 9 o'clock – we will go together, to the midnight mass, to hear the carols on the organs, and then I will take you back to your place . . .' You can hear the confident, almost imperious aristocratic tone in Émilie's offer. It makes a contrast to her usual self-deprecating comments about her lack of mathematical progress.

I presume she was referring to the famous Héloise, the tragic twelfth-century scholar and abbess who, as a teenager, had been the passionate student and lover of the philosopher Pierre Abelard. When their affair and subsequent secret marriage were discovered, Héloise's furious relatives castrated Abelard and the lovers were parted, seeking refuge in separate provincial monasteries. They corresponded throughout their lives, and were buried in the same grave. But whatever the identity of the girl or woman in Émilie's letter, Abelard's Héloise would have been one of few available role models for a passionate, intellectual woman like Émilie – especially a woman with a lover like Voltaire, whose *Philosophical Letters* was the cause of his current banishment from Paris, just as Abelard's rationalist philosophy had led to his twice being condemned for heresy and having to flee the authorities.

In the New Year, Émilie began studying mathematics with Clairaut, who would ultimately publish a textbook based on his lessons with her. He thought her ability 'altogether remarkable', which is high praise because he was a true mathematical prodigy. At the age of ten he was studying calculus, and at thirteen he presented an original paper on geometry at the Academy of Sciences; the Academy admitted him as a member when he was only eighteen. Now, in January 1735, he was still only twenty-one years old.

Two months later, the King's most trusted advisor, Cardinal Fleury, and his official, Chauvelin, reached the verdict that Voltaire had not intended to publish *Philosophical Letters* in France, and that it was his publisher alone who was guilty of publishing a book without official approval. Consequently, by the beginning of April, Émilie was able to write Sade the news they had all been waiting for: 'Voltaire has finally arrived [in Paris]; I believe the drama is over.' But Voltaire had received official permission to return to Paris only on the condition that he stopped writing outrageous things, and to guarantee that he did stop, the arrest warrant remained in force, a suspended sentence that could be reactivated at any time.

## *Chapter 4*

# ÉMILIE AND VOLTAIRE'S ACADEMY OF FREE THOUGHT

Voltaire was irrepressible. At Cirey, he had begun writing a bawdy poem about France's national heroine, Jeanne d'Arc (Joan of Arc), called *La Pucelle* (*The Virgin* or *The Maid*). It was a massive, pacy, racy, heroic-comic poem, full of sexual and other exploits, replete with literary and historical references and satirical allusions. Its continual, often crude emphasis on sex was used to highlight the stupidity of the leaders of the various factions, who were portrayed as being easily distracted from duty by their own carnal desires. There were even some lascivious descriptions of Joan herself, and smutty comments on her famed virginity. Irreverent, sexist, erudite but juvenile fun, it would seem.

However, it is interesting in this context that over the next few years, Voltaire and Émilie would become increasingly interested in deconstructing the myth-making that generally passed for history at that time. For centuries, history and religion had been linked together just as science and religion had been linked, so that both history and science traditionally had to fit in with religious doctrine and national myths. In the case of science, the rejection of a sun-centred cosmology on purely theological grounds – and the imprisonment of Galileo for believing in this heretical heliocentric cosmology – was a well-known example of the limiting effect of religion on scientific truth and freedom of thought. But Voltaire had noted in his *Letters* – and Émilie would note in her *Institutions de Physique* (*Fundamentals of Physics*) – that nationalism could have a similar effect.

In particular, they believed the French support for Descartes over Newton was partly influenced by nationalism, as was the British lionising of Newton at Descartes's and Leibniz's expense. As for history, Voltaire admired Jeanne d'Arc's courage and service to her country, but he did not like the way her story had been appropriated in the name of nationalism. He would later publish a passionate article in which he stripped from her legend every shred of religious honour and nationalistic glory; he described the farce of her trial and the horror of her immolation, concluding, 'One cannot conceive how we dare, after the countless horrors of which we have been guilty, call any [other] nation by the name of barbarian.'

This later work suggests there was a serious purpose in the licentious tone of *The Virgin* – namely, an attempt to challenge nationalistic and religious versions of history by humorously mocking the romanticism of Jeanne's legend. Whatever its intent, verses of Voltaire's ribald poem were now being savoured in Parisian salons, and, within six weeks of his release from 'house arrest' at Cirey, he had to flee Paris once again. Not that Voltaire had deliberately broken the law: as with *Philosophical Letters*, he had not formally published his poem; rather, handwritten verses had been circulated by one indiscreet friend after another, until *The Virgin* had become the talk of the town.

Émilie and her husband again negotiated on Voltaire's behalf, and eventually he was allowed once again to live under their 'supervision' at the château. Although he assumed the sequestration order would be removed soon enough, Voltaire decided to make Cirey his permanent base. Émilie remained in Paris, trying to maintain the public illusion that she was a faithful wife, and that she and her husband were merely Voltaire's protectors and she was simply his muse. But within days of his departure, she realised life had no meaning without love, and that Paris had 'become a desert' for her. She told Richelieu, 'I love Voltaire enough to sacrifice all that I could find pleasurable and agreeable in Paris for the happiness of living with him without danger, and the pleasure of tearing him away in spite of himself from his imprudence and his destiny.' A week later, on 15 June 1735, she left for Cirey.

Although love was uppermost in her mind, it was not emotional passion alone that led Émilie to abandon the Parisian lifestyle and move her household to Cirey: logic, too, played an important part in her decision. She had grown up in a world that assumed women were easily seduced by sensuality and romance, but which denied them real passion – a society that forced women into loveless arranged marriages and superficial affairs, but allowed them no active role in society, and few educational opportunities. Even the relatively enlightened Fontenelle believed women should be modest, dutiful and self-effacing, that they should 'conceal the knowledge acquired by their minds as much as [they should conceal] the natural sentiments of their hearts'. Some women were so unhappy they resorted to drugs like laudanum (a derivative of alcohol and opium), and many others developed psychosomatic illnesses.

By 'enlightened', I mean that Fontenelle had written on the need to separate faith and reason. In addition, at the age of twenty-nine, he had also pioneered the literary genre of popular science, by making the Cartesian view of astronomy accessible to the 'ladies' in his influential *Entretiens sur la pluralité des mondes* (*Conversations on the plurality of worlds*). *Conversations* was published in 1686, just a year before Newton's *Principia* first appeared. Fontenelle was a writer rather than a physicist, and he presented his material in a series of imaginary conversations between a debonair philosopher and a fictional marquise. Unlike Émilie, Fontenelle's marquise was not a scholar; rather, she was a curious and intelligent 'lady', and therefore the 'conversations' were tailored not for serious study but for interesting, light-hearted discussion in elegant Parisian salons.

Émilie, on the other hand, wanted the right to experience her full potential as a human being, to stretch her mind and express her heart with all her natural passion. As a woman, she was unable to enter the formal scientific academies, or to travel abroad to meet other scientists: not even the amiable marquis du Châtelet would permit such a thing, despite the fact that she dearly wanted to visit Britain to talk with Newton's disciples, as Voltaire and Maupertuis had done. But if she could not travel abroad to meet other scholars, then she would have them come to her: she and Voltaire would turn Cirey into both an internationally respected centre for intellectual creativity, and a true 'earthly paradise'.

At first, their friends and acquaintances found the idea of the enlightened couple at Cirey – living, working and loving together in freedom, unconstrained by the rules of court, religion or class – to be daring, fascinating, even transgressive. Visitors like Madame de Graffigny and Chevalier de Villefort gossiped in letters about the glamorous, eccentric lovers and their extraordinary way of life – their long hours of work punctuated by precisely scheduled coffee breaks, informal suppers and occasional formal soirées. Graffigny noted the couple seemed so much in love that sometimes they ate from the same spoon. However, Villefort told his correspondent, 'It was asked by the curious what the husband was doing all this time, but nobody knew.' Recall that 'the husband' spent little time at home because of his military duties; according to Graffigny, when he *was* in residence the marquis du Châtelet proved to be a quiet, retiring, rather boring member of the household, although he seemed happy enough during the occasional social evenings.

One of the first international scholars to make an extended stay was Francesco Algarotti. The son of a wealthy Venetian merchant, he was an intellectual adventurer who found a haven at Cirey, from where he told a friend, 'Here, far from the bustle of Paris, we lead lives full of intellectual pleasures . . . I am putting the last touches to my *Dialogues* [on Newton], which have found grace in the eyes of the belle Émilie and the savant Voltaire.' For his part, Voltaire said Algarotti 'makes verses like Ariosto, and knows his Locke and his Newton'. In a letter about life at Cirey during Algarotti's visit, Voltaire described how they would all read to each other from their works in progress, and then 'we return to Newton and to Locke, but not without drinking the wine of the country [Champagne] and enjoying excellent cheer, for we are very voluptuous philosophers'.

Madame de Graffigny concurred: having especially studied Locke's *Essay Concerning Human Understanding* in preparation for her visit, she was delighted with the life she found in that 'enchanted place'.

In a letter to her young friend François-Antoine Devaux, she enthusiastically described her first evening there, when everyone gathered together for supper: 'What did we not discuss? Poetry, sciences, the arts – all in a tone of badinage and good humour. How I wish I could reproduce them for you – these charming discourses, these enchanting discussions – but that is beyond me. The supper was not plentiful, but it was choice, tasty, and dainty; there was plenty of silver-plate on the table . . .'

Later, Graffigny described those special evenings when entertainment was arranged. For example, one or more of Voltaire's plays might be performed in Cirey's own little theatre, with everyone having their own part to play; one memorable day, they read or played thirty-three acts in twenty-four hours. Another night there were marionettes – Émilie had loved puppet shows as a child – with the dialogue being made up spontaneously, usually by Voltaire: he was 'as affable a child as he was sage a philosopher', according to Graffigny. On yet another night, Voltaire made everyone 'die of laughter', as Graffigny put it, with a magic lantern show in which he parodied his 'enemies' – namely, the literary critics who loved to hate him for his audacity and his success.

As for Émilie, the first thing Graffigny noticed about her was her rapid conversation: 'Her chatter is astonishing . . . [although] she speaks like an angel, I recognise that.' The next thing Graffigny noticed was her clothes: 'She wears an Indian gown and a large apron of black taffeta; her black hair is very long, and is fastened up at the back at the top of her head, and falls in ringlets like the hair of little children. It suits her very well.' Voltaire was different: 'I do not know whether he wore powder [and wig] on my account, but all I can say is that he makes as much display as though he were in Paris.'

At the famous Cirey soirées, Émilie, too, dressed as if she were 'in Paris', so her decision to dress down when she was working was significant: a rite of passage into a life of scholarship. It was such an unusual decision that even Voltaire once referred disdainfully to her ink-stained fingers and her old black apron. But dressing formally each day would have taken hours of precious time away from study, as she noted in the preface of her first independent work at Cirey, an expanded translation from English of Bernard Mandeville's book on ethics, *The Fable of the Bees*: 'Since I began to get to know myself, to pay attention to the price of time, to the brevity of life, to the uselessness of things of the world, I am astonished at having taken such extreme care of my teeth, my hair, but of having neglected my mind . . .' Now, at the age of twenty-eight, she was blissfully ensconced in her own 'academy' at Cirey, where she found herself 'very happy to have renounced, mid-stream, the frivolous things which occupy most women all their life . . .' Not that she gave up all frivolity and luxury: during her occasional sojourns at court she continued to win and lose enormous sums of money at cards, and she and Voltaire refurbished their rooms at Cirey in lavish rococo style.

Apart from the special soirées and the occasional days spent horse-riding with Émilie or hunting with Voltaire, visitors at Cirey were expected to spend long hours by themselves, because Émilie and Voltaire often worked all day and long into the night, breaking only for an hour or two's chat over coffee or supper. They read widely on science, philosophy and history: in their first two years at Cirey, they were especially interested in extending their ideas on liberty, free will, ethics, rationalist critiques of the Bible, and Newton – their thoughts on these seemingly disparate themes tumbling over each other, informing each other, making their discussions so exciting that Voltaire worried he was not writing enough, because too often 'the supper, Newton and Émilie carry me away'.

Nevertheless, he knew she was indispensable to his literary career, helping him with his research, reading and editing his work, and giving constructive criticism to such an extent that he told a friend, 'When Émilie is ill, I have no imagination.' For her part, she thought Voltaire was a 'universal man': a 'great metaphysician, a great historian, a great philosopher, [as well as the greatest French poet]'; he was her 'guide in the country of philosophy and reason'. He gave her a focus for exploring her own ideas, and to begin with, she helped him revise his *Traité de Métaphysique* (*Treatise on Metaphysics*). He also encouraged her to begin her own work, on *The Fable of the Bees.*

As well as using her 'translator's preface' as an opportunity to tell her story about renouncing the frivolous things of life in favour of study, Émilie also expressed in this preface her rage at the prejudice that prevented women from having access to a proper education. But the project was more than a feminist platform or a chance to improve her English. She was interested in *The Fable* for its own sake, too, and her shortened version of it – with her own additional comments at certain points – makes fascinating reading even today. Apart from its historical interest, it explores with both compassion and logic issues that are still contemporary, such as whether or not consumerism is a good thing for society; whether prostitution should be legalised; and the role of social conditioning in gender roles.

Mandeville's book had started life as a satirical verse poem, *The Grumbling Hive: or, Knaves turned Honest*. The poem's theme was that once society's 'do-gooders' had turned all the manipulative, greedy 'knaves' into 'virtuous' citizens, society crumbled for lack of enterprise. Mandeville believed societies are like beehives in which workers prosper through mutual self-interest rather than selflessness. The poem also showed the hypocrisy of those who claimed only the 'lower' classes succumed to pride, greed, and other 'sins', and in subsequent editions over thirty years (the latest being published in 1734), Mandeville added a long essay and explanatory comments; he called this expanded book, *The Fable of the Bees: or, Private vices, Publick benefits*. Émilie's version was based on Mandeville's essay 'An Enquiry into the Origin of Moral Virtue', and on some of his additional comments. In particular, she emphasised the section on the socially imposed relationship between gender and virtue.

In fact, in both Britain and France, one of the most controversial aspects of *The Fable* was Mandeville's approach to the very notion of 'virtue', and its opposite, 'vice', and Émilie was impressed that he approached the topic by trying to understand human nature in a rational rather than an emotive way. He concluded it is impossible to ban 'vice'; instead we should develop laws that tolerate the lesser vices in order to prevent worse ones. For example, he argued that regulated prostitution helped prevent violent sexual crimes against women, and that commerce made society prosperous even though it is built largely on the vices of greed, dishonesty and consumerism (which Émilie called 'love of luxury').

One of the longest sections of Émilie's version of the book is on honour and shame, particularly as it affects women. In Émilie's words, Mandeville said, 'To continually declaim against [debauchery] is not to know men'; in particular, he felt the moralists often mistook 'modesty' for 'virtue': he argued, for example, that a little girl will blush if she inadvertently shows her leg in public, but this is not natural modesty – rather, it has been taught her from infancy, long before she is capable of understanding its purpose, which is to ensure chastity. 'The difference in modesty between the sexes is often attributed to nature, but it is only the effect of [those] first instructions,' he said. He also described the tragedy that follows for a good woman – 'perhaps even a religious woman if you want' – who yields to temptation and becomes pregnant before she is married: she will lose her reputation, her self-esteem, and her chance for a good marriage. His point was that obeying social custom is not the same as being moral: the 'fallen' woman is scorned but no-one judges the man who rapes his 'lawful' bride.

During an extended visit at Cirey several years later, Madame de Graffigny would read the manuscript of Émilie's French version of *The Fable*, and would write Devaux, 'What a woman she is! How small I am compared to her! If I were diminished physically in proportion, I should be able to escape through the keyhole.' Referring to the clarity of Émilie's thought, she added, 'It is true that when women meddle with writing, they surpass men . . . But how many centuries does it take to produce a woman like her?'

Françoise de Graffigny herself was an unusual woman, and she would later use what she learned at Cirey in her own writing career: at the time of her visit in late 1738, she was just beginning to reinvent herself as a writer, having recently left her violent, abusive husband, and having lost her five children, who had all died as infants. Voltaire's play, *Alzire*, and Émilie's version of *The Fable*, would inspire Graffigny's later novel, *Lettres d'une Peruvienne* (*Peruvian Letters*). *Alzire* had used Peru as an exotic location to explore the meaning of 'natural virtue' in the context of religious tolerance. It was set during the sixteenth-century Spanish conquest of Peru, and it aimed to show that ethics, or 'virtue', was based on natural human decency rather than on slavish adherence to religious ritual, pagan or Christian; in other words, it aimed to show it was possible to be a good person without the aid of religious dogma. Émilie's *Fable* had analysed 'virtue' in a similar but broader context, with an emphasis on gender conditioning and sexual stereotypes. Now Graffigny wanted to explore this idea in relation to the sexual double standard, in which 'virtue' meant one thing for women – being faithful, or at least discreet, wives – and quite another for men, who were allowed their dalliances, so that virtue for them meant courage, honour, rationality. Émilie no doubt provided the model for Graffigny's free-spirited Peruvian heroine, Zilia, who wants a life of independence – a life she realises is not considered proper for women in France. *Peruvian Letters*, published in 1747, became one of the most popular novels of the century.

Émilie had begun work on *The Fable* with no thought of immediate publication, because the English edition had already been officially proscribed and burned in France. Instead, she was motivated simply by the spirit of scholarship: her work on *The Fable* was a chance to integrate and critique the ideas not only of Mandeville, but also of Locke, Voltaire, and others who had written on the still-contested concept of 'secular ethics'. For example, she disagreed with Locke on the existence of innate ideas. Locke had rejected the Cartesian (and Platonic) doctrine that we are born with a predisposition to certain innate concepts and moral values, and had argued instead that experience is the primary source of all our ideas, so that God does not plant morality into our souls, we have to make our own moral rules. Leibniz had objected to Locke's empiricism, saying that surely abstract mathematical ideas do not arise from everyday experience. (In this respect, the status of mathematics is still a tantalising question, although great philosophers have been discussing it for over two thousand years.) Émilie's objection to Locke's view concerned morality rather than mathematics: she believed some moral precepts – like the 'golden rule' of treating others as you would want them to treat you – were so universal they must be innate. In this instance, she had missed Locke's point – namely, that the universality of certain ideas does not *prove* they are innate – but she can be forgiven because Locke's long and brilliant *Essay* was sometimes rambling and confusing. Besides, most of her work on *The Fable* was insightful and incisive, and the process of writing it proved to be excellent intellectual training for all her subsequent projects, beginning with a joint venture with Voltaire on Newton himself.

Both Émilie and Voltaire tended to work on several topics at once, and, in January 1736, she wrote to her friend Sade: 'We are going to play, in our little Republic of Cirey, a comedy Voltaire wrote for us . . . He is also writing the history of Louis XIV, and me, I am Newtonising.' She also told Sade about Algarotti: 'He is a young Venetian [who] has been putting Newton's sublime discoveries on the nature of light into a dialogue, in the style of the [Cartesian] *Conversations* of Fontenelle.' These 'sublime' discoveries included the fact that ordinary 'white' light is made up of the colours of the rainbow – it was Newton who chose the word 'spectrum' (from the Latin for 'image' or 'apparition') to describe his discovery. Émilie was studying his book *Opticks* as part of her 'Newtonising' at Cirey.

In contrast to the *Principia*, the strength of *Opticks* lay in its descriptions of Newton's groundbreaking experiments rather than in mathematical theorems: it would take another two centuries for someone to do for light what Newton did for gravity – that is, to express its fundamental nature in a rigorous and elegant mathematical theory. After all, Newton had built his theory of gravity on the pioneering experiments and observations of Galileo and Kepler, so when it came to the study of light, he had to play the role of Galileo. It was such an important role that Émilie and Voltaire wanted to bring both *Opticks* and the *Principia* to a wider French audience. Their proposed book – *Éléments de la philosophie de Neuton* (*Elements of Newton's Philosophy*) – would prove to be a milestone in popular Continental Newtonianism.

Clearly, they had been inspired in this venture by Algarotti's work at Cirey. But he had taken a light-hearted approach to the more accessible aspects of Newton's work – namely, the experiments on light and colour. His book was called *Newtonianism for the Ladies, or Dialogues on light and colours*, and he had used Fontenelle's device of a flirtatious, fictitious philosopher/narrator 'explaining' scientific concepts to a curious, intelligent 'marquise', although Algarotti's marquise was also somewhat flighty and easily distracted. Émilie and Voltaire were interested in a more serious style of science writing, and they wanted to include the more difficult aspects of Newton's philosophy.

The first significant 'popularisations' of the more challenging, gravitational concepts in Newton's new science had been made by his own friends and colleagues: in particular, Locke's twenty-five-page pamphlet, *Elements of Natural Philosophy*, published in 1720, and Henry Pemberton's *View of Sir Isaac Newton's Philosophy* of 1728. (Newton had personally chosen Pemberton to edit his third and final edition of the *Principia* – the edition that Émilie would translate.) John Keill in Britain and Willem 's-Gravesande in Holland also produced influential books but, like Pemberton's work, they were at quite an advanced level. Émilie and Voltaire aimed for the middle ground between these books and Algarotti's, and their *Elements of Newton's Philosophy* was the next major popularisation of gravitational theory, and it was also the first French one (apart from Maupertuis's semi-popular commentary in his *Discourse* of 1732, and Voltaire's nine-page chapter on gravity in *Philosophical Letters*).

Voltaire was the chief author of *Elements of Newton's Philosophy*, and his name alone would appear on the cover, but Émilie was his advisor, co-researcher and inspiration: she made copious notes on various topics, some of which Voltaire used word for word. Consequently, she is entitled to be seen as co-author of the book, and Voltaire himself regarded her in this way: he began the book with a preface that singled her out as both muse and colleague. He also added a dedicatory poem in which he referred to her as Minerva, the Roman goddess of wisdom: 'You call me to you, vast and powerful Genius, Minerva of France, immortal Émilie, disciple of Newton and of Truth.' Privately, he was even more direct about Émilie's role in producing the book:

he told his young Francophile friend Crown Prince Frederick of Prussia, 'Minerva dictated and I wrote.'

Much of the *Elements* discussed the history and philosophy of science, giving an overview of the development of scientific thinking on light and gravity, from Aristotle to Newton. In particular, Émilie and Voltaire took care to highlight not only Descartes's outmoded ideas – the celestial vortices, and the idea that light exists as tiny fragments of dust left over from the primeval 'cubes' out of which God created the planets, oceans and the air – but also the more 'modern' work of Kepler, Galileo and others on whom Newton drew. But although the *Elements* was designed to convince Cartesians to give up their allegiance to Descartes's old-fashioned ideas, it did acknowledge the importance of Descartes as an influence on Newton.

This acknowledgment is important, because the emphasis on Newton in Émilie and Voltaire's story – as well as in scientific history today – can give the mistaken impression that Descartes was a mediocre talent, and that therefore Continental Cartesians were simplistically naive or wilful in their adherence to his ideas. But as Voltaire had already implied in his *Philosophical Letters*, it was Descartes who laid the foundations for some of Newton's laws of motion, and he is the most important forerunner of modern 'analytic geometry', the mathematical method by which geometrical curves like circles and ellipses are analysed in terms of algebraic relationships between the coordinates of points lying on the curves. This 'algebraic geometry' is vital because it underlies calculus, which is the language of modern physics. The familiar $x$ and $y$ coordinates of school textbooks are called 'Cartesian coordinates' in Descartes's honour, and it was in large part through his work that modern mathematics began to develop as a language in its own right.

As for Descartes's philosophy, Émilie, in particular, agreed he had made a great advance with his method of reasoning via 'clear conceptualisation' and 'systematic doubt' (which she would discuss in more detail in a work of her own, *Fundamentals of Physics*). This Cartesian method of arriving at 'truth' was a deductive method based on scepticism: if something can be doubted, we cannot assume it is true, and so he used arguments such as, 'If I am thinking, then I must exist, and this can never be doubted as long as I am thinking'. (This is the key to the famous line, 'I think therefore I am'.) In relation to 'clear conceptualisation', he believed that if we can clearly conceive of something, then it is 'true', because he believed the intellect – in contrast to the imagination and to everyday intuition – cannot be deceived, whereas our notions of perceived everyday reality may be inaccurate because of the limitations of our physical senses. At this point, according to Émilie, Descartes pushed his method too far, because he did not take sufficient account of experiment. It was a matter of rationalism versus empiricism, on which Aristotle had challenged Plato himself, just as Newton and Locke challenged Descartes.

Because of its lack of experimental foundation, the Cartesian method is circular when it is applied to theoretical physics. Recall the Cartesians believed theories should be created from hypothetical but clearly articulated mechanical *causes*. When it came to a theory of planetary motion, they found it easy to conceive the cause of this motion in terms of the impetuses of cosmic vortices, but they could not conceptualise a disembodied, remote force of gravity. But the circularity arises because in the Cartesian method, cause precedes effect – a logical enough point of view on the face of it. In terms of making theories, however, this meant possible effects were to be theoretically *deduced* from the assumed causes, and then compared with what happens in nature; such a causal hypothesis passed for a theory as long as it remained uncontradicted by physical facts. Newton's approach proceeded in the opposite way, as Émilie and Voltaire pointed out. His method began not with causes but with effects – in particular, mathematical descriptions of experimentally observed facts (such as Kepler's analyses of planetary motion and Galileo's observations of the effects of gravity). Then, by purely logical inferences or mathematical deductions from these descriptions, new equations were discovered, such as the inverse-square law. Newton's method is by far the more rigorous, because it assumes only known facts. By contrast, Descartes's method allowed the simple fact of planetary motion to 'prove' the existence of his purely hypothetical celestial vortices.

Consequently, Émilie and Voltaire told their readers that, using the Cartesian approach, a theory could, at best, uncover the truth purely by chance – by a chance match between physical facts and the initial speculative hypothesis – which is why the *Elements* set out to discuss only the *effects* of gravity, not its possible origins: '[I]f we see its effects, we know it exists; we must *not* begin by imagining the cause and then making hypotheses, because that is the surest way of losing our way. Instead, let us follow step by step what actually happens in Nature: like voyagers who have arrived at the mouth of a river, we must travel up the river before imagining where its source is located.'

*Elements of Newton's Philosophy* helped pioneer a bold new genre of serious but relatively accessible science writing by attempting to explain a wide range of mathematical and scientific phenomena. Without Émilie's mathematical assistance, however, it is unlikely Voltaire would have embarked on such a challenging project. He stated in his preface that his goal was to make Newton's work accessible to his readers 'without effort' on their part – especially readers 'who had heard of Newton and his philosophy by name only'. But Voltaire himself had expended a great deal of effort trying to understand Newton's works: 'I am straining my brain with Newton', he wrote to a friend at this time. In another letter, he spoke of 'a devil of a Newton, who finds precisely how much the sun weighs, and which colours light is composed of. This strange man turns my head.' And, in a comment that proves just how much he needed Émilie, he wrote to another friend, 'I have health too frail to apply myself to mathematics. I cannot work more than an hour a day without a lot of suffering.'

One of the more technical, mathematical sections of the *Elements* concerned Kepler's observationally derived astronomical laws, on which Newton had built his theory of gravity. Kepler's first law is the observed fact that all the planets have elliptical orbits around the sun; his second law revealed that each planet's speed varies at different points of its orbit – that is, at different times of the year. Kepler discovered this from his brilliant geometrical analysis of the imaginary wedges or 'sectors' swept out in a given time by the planets as they move in their elliptical orbits around the sun. As Émilie and Voltaire noted, however, it was Newton's incomparable genius that uncovered the true import of Kepler's second law.

To get a feel for this seminal result, imagine the minute hand on a circular clock. When the hand moves between, say, ten and twenty minutes past the hour, you can visualise it as sweeping out a ten-minute 'sector' or wedge of the circle. The key point of this analogy is that the sector swept out in *any* ten-minute interval will have the same shape and size, no matter where on the clock the ten-minute sector is taken – and so any sector swept out in the same period of time will have the same area. (This is illustrated in the left-hand diagram in Figure 1 of the Appendix.) In Kepler's second law, the idea is to adapt this image to the elliptical 'sectors' swept out by the planets, but as you can see in the right-hand diagram in Figure 1 (and the *Elements* contained a similar diagram), the shapes of these planetary sectors vary with position: when the planet is further away from the sun, the sectors have long 'hands' and short orbital arcs, while closer to the sun, the 'hands' are shorter but the orbital arc is longer. But Kepler discovered a fascinating fact of nature: from his observational data on planetary positions throughout the year, he noticed that although two different sectors swept out in the same time may have different shapes, they still have the same *areas*. This means the planet needs to move faster when it is closer to the sun, so it can cover the longer arc of orbit in the same time as the shorter arc. Émilie and Voltaire pointed out that this made sense because when the planet is closest to the sun, it feels the most gravity, which causes it to move more quickly.

However, this is not the reason that Kepler's observationally derived 'law of equal areas' is so important. As Émilie and Voltaire noted, one of Newton's great insights was his realisation that Kepler's second law was a consequence of the force of gravity that caused the planets to move about the sun, and that this law therefore defined the very notion of gravitational force. Newton's inverse-square law described the effective *strength* of gravity at any point, but the 'law of equal areas' gave a conceptual framework for describing force itself – or more precisely, for describing the behaviour over time of any attractive, inward-pointing force that causes a body to move towards or around a fixed 'centre', a force that Newton therefore called 'centripetal'.

The significance of Newton's definition is apparent when you compare it with the traditional notion of force as an impulse, a single direct impetus or series of impetuses rather than something acting continuously over time, like gravity. Newton's use of the law of equal areas provided a way of conceptualising and quantifying continuous orbital forces (in terms of what modern physicists call 'conservation of angular momentum'): if equal sectors or angular areas are swept out in equal times, then the ratio of area to time does not change – if you double the time, you double the area, as Émilie and Voltaire pointed out. (This means the *rate* at which the object is sweeping out an angular area is constant throughout the motion – that is, it is 'conserved'.)

Centripetal force also differed from Huygens's earlier concept of 'centrifugal' force. This was assumed to be an outward-directed force such as you might imagine when you whirl a stone that is tied to a string: if the string breaks, the stone flies off in an outward direction, so Huygens assumed the force propelling it must also be outward. It was Robert Hooke who suggested to Newton the idea of an inward-pointing force in orbital motion, but it was Newton who provided the mathematical framework: rather than focusing on the tendency of a whirling stone to fly off when the string is *broken*, he provided a mathematical analysis of the force that actually keeps the object in orbit, namely the centripetal force, which Voltaire and Émilie called the 'true' force. Indeed, centrifugal force is often called a 'pseudo-force' today, because it can be viewed as an artefact of 'pre-existing' rotatory motion (or, technically speaking, an artefact of the choice of an 'accelerated frame of reference'), whereas 'real' forces are considered to be those attributable to a natural phenomenon – specifically gravity, electromagnetism or the strong and weak nuclear forces in atomic nuclei. Newton showed that the planets' orbital motion was governed by the centripetal force of gravity; similarly, instead of speaking of centrifugal force, physicists would say the whirling stone is governed by an inward-pointing centripetal force – namely, the tension in the string, which is caused by the electromagnetic forces between the atoms of the string. These interatomic electromagnetic forces are very strong in solid objects – it is these forces that bind the atoms together and keep the objects solid; hence electromagnetism is also the source of the everyday 'push-pull forces' allowed by the Cartesians, although the electrical structure of atoms was unknown in the seventeenth and eighteenth centuries.

Ultimately, though, scientific definitions of reality are a matter of language and philosophy, not of objective truth. Consequently, although Newton's framework gives the clearest conceptual basis for analysing motion in everyday physics, centrifugal force is still a useful concept in certain situations. In fact, Newton himself made startling use of it, as the next chapter shows. Incidentally, by 'everyday physics', I mean that which describes the reality we intuitively see around us: the motions of planets, cricket balls, aeroplanes, falling apples, and so on. By contrast, Einsteinian relativity comes into its own when things are extremely fast (like electrons or light photons) or extremely massive (like stars and galaxies), and quantum theory comes into its own when things are extremely small (such as subatomic particles, although the *effects* of quantum processes may be felt over large distances). In this book, I am focusing mainly on everyday physics, because relativity and quantum theory were, of course, unknown in both Émilie's time and Mary Somerville's. (However, to illustrate the point about scientific definitions being a matter of language and philosophy, in Einstein's general relativity theory the role of frames of reference differs from that in Newtonian physics: 'pseudo-forces' are called 'inertial forces' and they are defined analogously to 'real' gravitational force.)

The *Elements* gave a detailed outline of Newton's mathematical proof of the 'law of equal areas' as applied to centripetal force: it uses high-school-level geometry to find the areas of various triangles that approximate the sectors swept out in various times. Continuous motion is simulated by assuming the total area swept out by an orbiting body is made up of an infinite number of *infinitesimal* triangles, each denoting the infinitesimal area swept out in an 'instant' of time. (The use of infinitesimals is at the heart of calculus.)

Émilie and Voltaire also discussed Kepler's third and final law, which links the length of a planet's year with the radius of its orbit. It would seem natural to suppose that the greater the radius ($r$), the longer the distance travelled in one orbit and the longer the time ($T$) taken to make a single round trip (that is, the longer the planet's year). But Kepler found the relationship is not simply that $T$ is proportional to $r$; rather, $T^2$ is proportional to $r^3$. In fact, the proportionality factor itself is the same for every planet, but Émilie and Voltaire wisely noted it would be easiest for their readers to understand the significance of Kepler's 'admirable rule' through an example. They showed that by comparing the orbital times of the earth and another planet, such as Venus, and using the known distance between the earth and the sun, Kepler's third law can be used to calculate the distance between Venus and the sun. It is an extraordinary result: as long as you know the length of each planet's 'year', and as long as you *also* know the distance to the sun for a reference planet such as the earth, then the distances for all the other planets can be calculated from Kepler's law, using elementary high-school algebra rather than expensive, difficult astronomical measurements. Kepler's third law can also be applied, with a different proportionality factor, to all the satellites of a given planet – and in the Appendix, you can see the example given in the *Elements*, namely, how to calculate the distances of Jupiter's moons.

Interestingly, Émilie and Voltaire noted that Kepler was not as good a philosopher as he was an astronomer, because he could not explain his results; instead, he supposed the sun contained some sort of vegetable 'soul', and that it drew in the planets by centrifugal force as it turned on its own axis. Furthermore, according to the *Elements*, Kepler's explanation for why the planets did not fall into the sun was that they, too, were rotating, in such a way that they presented alternately 'hostile' and 'friendly' faces to the sun, the friendly face being drawn in and the hostile one being repulsed. This series of mysterious attractions and repulsions supposedly kept the planets in orbit. As Émilie and Voltaire pointed out, it was an incredible contrast with Newton's explanation of Kepler's laws in terms of the centripetal force of gravity.

They also pointed out that Newton used Kepler's third law to disprove Descartes's idea that the planets were moved by ethereal vortices: the mathematical analysis of such a vortex shows that the planetary orbits would *not* obey Kepler's third law, and since this law was based on physical fact, the assumption of planetary vortices must be wrong. (It was a brilliant analysis, and the conclusion was correct, although modern critics have pointed out flaws in Newton's proof.) The *Elements* also mentioned Newton's argument that vortices are inconsistent with the gravitational theory of comets: comets travel across the orbital plane of planets from such great distances and at such huge speeds that their massive swirling vortices would disrupt the planetary motions, but this is contrary to observation.

Despite the profundity and reasonableness of Newton's method of creating a theory by working 'backwards' – by beginning with what actually happens rather than by trying to second-guess nature at the outset by *assuming* specific causes for natural phenomena – nevertheless, when Émilie and Voltaire were writing their *Elements*, Newton's theory of gravity was far from secure. It was not simply the philosophical problem of action-at-a-distance – the lack of an understandable mechanism by which gravity itself acted – it was also a matter of experimental verification and the fact that, for all its marvellous successes, the inverse-square law did not yet account for all the known facts of celestial motion. Newton himself was aware of this, particularly with respect to the moon's motion (of which more later). Consequently, there was still much work to be done in testing and developing Newton's ideas. In fact, unbeknown to Émilie and Voltaire, Maupertuis was already preparing such a test.

## Chapter 5

# TESTING NEWTON: THE 'NEW ARGONAUTS'

While working on the *Elements*, Émilie had written to Maupertuis, 'in the hope of drawing you to Cirey', that she and Voltaire had 'a fine physics laboratory, telescopes, and mountains from where one can enjoy a vast horizon'. However, she soon discovered Maupertuis was caught up in a Newtonian adventure of his own. He was preparing to lead a scientific expedition to the Arctic, in order to take measurements that would determine whether the earth really was elongated at the poles – as found by measurements made in France by Jacques Cassini – or whether it was flattened at the poles as Newton had predicted. On hearing Maupertuis's news, Émilie wrote him in her half-joking, half-serious way, 'You are not content with abandoning me for the pole, you are also stealing Clairaut from me . . . Adieu, Monsieur. Voltaire and I will drink your health with the wine of Alicante [in south-east Spain].'

Determining the true shape of the earth was important scientifically, because it would provide physical confirmation – or otherwise – of Newton's theory of gravity. But it was for practical reasons that, through the auspices of the Academy of Sciences, the French king had funded first Cassini's work, and now Maupertuis's prospective expedition, together with another one to Mitad del Mundo, which is now in Ecuador but was then part of Peru. (The Peruvian expedition was led by Voltaire's old friend La Condamine. As the inimitable Émilie told Maupertuis, 'You are going to freeze for glory, while La Condamine will boil.') Accurate knowledge of the shape of the earth has relevance to cartography and navigation, because, as Maupertuis put it (with slight exaggeration), 'In a course of one hundred degrees longitude, there might be a mistake of more than two degrees, if sailing really upon Sir Isaac Newton's Earth one should imagine himself to be upon Mr Cassini's. And how many ships have perished by smaller mistakes?'

It turns out that if the earth is flattened at the poles, then the north-south distance along a meridian of longitude – measured between two lines of latitude that are, say, one degree apart – will be greater if those lines are at polar latitudes rather than at equatorial ones. This is because latitude itself is defined as if the earth were spherical, and in the Appendix I have given a little more explanation of how this relates to the 'flattened' north–south distance. But the important thing to note here is that Maupertuis's Arctic base was about sixty-six degrees north of the equator and La Condamine's equatorial site was about one degree south of the equator – and their respective measurements would show whether or not the measured north-south distance in travelling between two lines of latitude at sixty-six and sixty-seven degrees was, indeed, longer than the distance between two lines at around one and two degrees.

The remote and dangerous locations of these two expeditions showed science in a daring and glamorous light. When formally announcing the departure of La Condamine's team, at a public meeting in April 1735, Fontenelle had asked dramatically, 'How many hardships, and fearful hardships, must accompany such an enterprise? How many unforeseen perils? And what glory must surely redound to the new Argonauts?' For Maupertuis, however, 'This voyage would hardly suit me if I were content and happy.' Émilie had already noticed his unhappiness when she visited him and Clairaut at their retreat at Mont Valérien, west of Paris, and she wrote to Richelieu, 'He has a restlessness of spirit which makes him very unhappy and which proves that it is more important to occupy his heart than his mind. But unfortunately it is easier to do algebra than to be in love.'

Maupertuis did, indeed, throw himself into 'algebra', in order to understand and develop Newton's mathematics of the shape of the earth. Publicly, though, he exaggerated the practical importance of the expedition, to hide the fact that he was more interested in checking the mathematics than in navigation and map-making. Despite his dissembling, the fact that he was a Newtonian and Cassini was a Cartesian made many perceive the Arctic venture as a contest between Newton and Descartes, although Descartes had not made any prediction about the shape of the earth. However, Huygens, the brilliant neo-Cartesian, *had* made such a prediction: unlike Cassini, Huygens had predicted a flattening rather than an elongation of the earth, but the amount of flattening he calculated was different from Newton's.

Newton argued – and Huygens had independently offered a similar argument – that the earth was essentially fluid (its surface is mostly sea, and its core is molten); consequently, as the earth rotates on its axis, its fluid bulges at the equator because of the centrifugal 'force' of its spinning motion. To see this, take a ball or an orange and rotate it around its vertical axis: the top and bottom points of the object – its poles – simply turn around in a circle on the spot, but a point marked on the middle of the object – on its 'equator' – will travel in a circle the size of the equator itself. So, in one twenty-four-hour rotation of the earth, points at latitudes near the poles travel in small circles, while points near the equator travel a much larger distance in the same time. Which means particles of fluid near the equator must be travelling at much greater speeds than those near the poles, and so they have a greater tendency to 'fly off' the earth as it spins around. Recall that this tendency to 'fly off' the path is what is meant by centrifugal force.

Of course, the fluid particles on the earth's surface are not actually flung off the globe, because we are also 'particles' on the earth's surface and we are not thrown into space. Clearly, then, the force of gravity that keeps us 'attached' to the ground is much stronger than the centrifugal force of the earth's rotation. Both Newton and Huygens agreed that the *net* force keeping us on the earth is the force of gravity minus the (radial component of the) centrifugal force, and that this net force will be *smallest at the equator*, because the centrifugal force there is greatest.

This 'centrifugal force' argument was supported by a surprising fact: the net downward gravitational force does, indeed, vary with latitude – it is least at the equator and greatest at the poles. Jean Richer had discovered this extraordinary fact in 1672, during an expedition to Cayenne (the capital of French Guiana, where he was making astronomical measurements that would enable more accurate calculations of the distance between the earth and the sun). Richer noticed that his pendulum clock ran slower in Cayenne than in Paris: that is, the duration of one day – as measured by the earth's rotation with respect to the stars – was nearly two and a half minutes less by his clock at Cayenne than the accepted measurement made in Paris.

Now, Huygens had already discovered that the period of such a pendulum – that is, the time between successive swings or beats – depends only on the length of the pendulum and on the net force of gravity; in fact, it turns out – as you can see from the formula in the Appendix – that the longer the pendulum and the less the gravity, the larger the period, and therefore the slower the clock ticks. Describing Richer's discovery in the *Principia*, Newton noted that changes in temperature from one latitude to another would not substantially affect the length of a clock's pendulum (as Newton had shown from his own experiments), so it must be the net force of gravity that had changed, in such a way that it was less at the equator than in Paris, causing Richer's clock to run slow. Such a change of net gravity fitted in beautifully with the centrifugal force argument, as Huygens had been the first to realise.

Although Huygens had independently suggested the earth was oblate, and although he was the first to work out the ratio of gravity to centrifugal force in the context of the earth's rotation, he had not used his results to make any quantitative predictions about the consequent amount of flattening at the poles of the earth until *after* he read the *Principia*. He was so inspired by Newton's work on this topic that he used Newton's method to recalculate the shape of the earth in terms of a neo-Cartesian idea of gravity, in which impulsions in some kind of invisible ethereal matter produced the downward pull of gravity. Cassini's measurements notwithstanding, then, it seemed very likely that the earth was flattened rather than elongated at the poles. But the crucial question was, how much was it flattened? The calculated prediction depended on how much gravity there was to counteract the centrifugal force at various points on the earth's surface – and *that* depended on which theory of gravity you used. It was here that Huygens parted company with Newton.

Newton's ingenious approach was based on an amazingly simple 'thought experiment' that shows the extraordinarily intuitive way in which he worked. First, imagine the earth is a perfect sphere, and imagine a 'north–south' channel (or tunnel) dug out along its radius from the North Pole to the centre of the earth; similarly, imagine an 'east–west' channel of the same width, dug along the radius from a point on the earth's surface at the equator through to the centre of the earth. The radii are both equal, because we are considering a perfect sphere, so the two channels are of equal width and length, and when filled with water, they should each weigh the same. (In reality, these channels are part of the earth itself, but Newton imagined them filled with water because his calculations relied on the simplifying assumption that, like water, the earth's matter was homogeneous throughout both channels.) But weight varies with latitude, as the pendulum experiments showed (because 'weight' is simply the net downward force of gravity acting on a body). Taking this observational fact alone – without assuming it is due to centrifugal force, which in turn affects the shape of the earth, and without yet considering the law of gravity itself – then, in a perfectly spherical earth, the east–west (equatorial) column of water would weigh *less* than the north–south one, because net gravity is least near the equator.

Newton argued the two columns of fluid should actually weigh the *same*, in order not to upset the earth's equilibrium: if the earth were motionless, and if it were heavier in one direction than in the other, then it would tend to roll towards one side rather than being in equilibrium. In order to keep the earth in balance, both columns of fluid should weigh the same, and so the equatorial column must be *longer*, so that it can contain more water in order to weigh the same as the intrinsically heavier north–south column. If the equatorial radius is longer than the north–south radius, then the earth must bulge in the middle and be flattened at the poles.

Newton did not stop there: he was not content to give only a qualitative argument but actually made a quantitative prediction, by calculating the extra length needed for the equatorial radius; he did this by calculating the extra weight needed from his equation for the centrifugal forces at various latitudes and from his formula for the force of gravity. Huygens did the same (he had borrowed Newton's 'channel' argument), but to calculate the extra weight needed, he used a much more simplistic, neo-Cartesian theory of gravity. The difference between these theories is yet further evidence of Newton's rare genius.

Unlike Huygens and the other Cartesians, Newton assumed every particle of matter has gravity – it is not just a property of huge bodies like the earth and the sun. It seemed a strange idea, but Émilie would later explain it well, in her commentary on the *Principia*. She said the total gravitational force of a planet must be composed from the forces of its parts, because 'if one imagines that several small planets united to form a large one, the gravitational force of this large planet would be composed of the forces of all the small planets' (or small particles of matter). In other words, in Newton's theory, each particle of matter exerts a gravitational force on any other particle, and the strength of this force is not only inversely proportional to the square of the distance between the particles, it is also proportional to their masses. But Newton showed that the gravitational effect of an object depends on its *shape* as well as its mass: Émilie explained that in one of a group of theorems now justly called the 'superb' theorems, he proved mathematically that his formula for gravity applies to *separate*, *spherical* particles, whether they are huge planets or tiny 'atoms' of matter. It is the law of attraction between one spherical body and another.

He also proved that these spherical objects attract each other as if all their mass is concentrated at their centres, so that in applying the inverse-square law, distance is measured from one centre to the other. But the catch arose when you wanted to find the gravity at any point *inside* such an object – inside the earth, say, in order to work out the weight of a column of fluid in the earth. In such a case, you had to find the *total effect* of the *separate* inverse-square forces exerted at any point by each of the component particles of matter, and in one of his 'superb' theorems, Newton showed this total effect did *not* add up to an overall inverse-square law. He found an equivalent result for the gravity inside a flattened sphere (and he found the forces of gravity inside and outside other variously shaped objects). I will return to these theorems later, when I talk more about Émilie's work on the *Principia*. For the moment, it is enough to note that Maupertuis had 'tortured' himself, as he put it, trying to understand them – and that Huygens had assumed the earth's gravity was independent of its mass and shape when he made his own predictions of the shape of the earth.

By the summer of 1736, Maupertuis and Clairaut were in Lapland, home of the Sami people. They were based at Tornio, in present-day Finland, right on the border with Sweden – but at the time, the area was Swedish and the village was generally known as Torneå. The Tornio River runs in a roughly north–south direction, so the distance along a meridian could be measured by travelling north up the river valley.

Maupertuis's scientific team included the Swedish astronomer Anders Celsius, who is famous today for his centigrade or 'Celsius' temperature scale. He had visited Maupertuis and Clairaut at Mont Valérien, where they had enthusiastically discussed the idea of the expedition. The bond of friendship between these scientists, together with Maupertuis's decisive leadership, proved crucial to the success of the venture. By contrast, La Condamine's scientific colleagues undermined his authority, and he had to deal with all the unforeseen disasters that dogged the Peruvian expedition – including harassment from some of the local authorities, the murder of the team's surgeon, the effects of tropical diseases, and lack of interest in the project from untrained local workers and also, after months and years in the unbearable heat, from some of the French technical crew itself. Maupertuis had luck on his side, but the involvement of Celsius certainly helped to smooth the way with the Swedish authorities, and the king himself provided a regiment of soldiers to assist with the labouring work of the expedition.

In addition to the scientists and soldiers, there was also a support crew, including a Finnish navigator and interpreter, Anders Hellant, and a number of oarsmen: in measuring the north–south distance along the ground between two lines of latitude one degree apart, the Arctic Argonauts had to travel about one hundred kilometres up the river valley, and part of the journey was made by boat, navigating dangerous rapids in the process. Maupertuis later vividly recounted that 'in the middle of a cataract, whose noise was terrifying', the 'frail' boat – 'carried by a torrent of waves, foam and stones, sometimes tossed in the air, sometimes lost in the torrents' – was expertly steered by three 'intrepid Finns'. It was a 'spectacular' experience.

Unfortunately, much of the route did not follow the river, and the journey had to be made on foot, through dense forests and insect-infested swamps. At one point the party met some reindeer herders who told them they could deter the mosquitoes by sitting in the smoke of their campfire, but Maupertuis would later recall that soon they came upon insects so 'cruel' that smoke did not affect them: 'The insects infected whatever we wanted to eat; our food was constantly black with them.'

The team used the stars to determine their latitudes, but to measure the longitudinal distance along the ground, they used the method of triangulation that is used by surveyors. A geophysical triangle can be created by marking three spots on the landscape – two on the ground and the third at a high point like a mountaintop; these three points define the vertices of the triangle, and in this way, Maupertuis's team created a network of contiguous triangles all along the valley. In geophysical contexts, angles are much easier to measure than lengths, and triangulation makes use of this because it is based on the fact that if you can measure the length of one side of a triangle, and you can measure two of its angles, then you can use simple high-school trigonometry to calculate the lengths of its other two sides (as you can see in the Appendix). These two sides are also part of the two adjacent triangles in the contiguous network, so for these next two triangles, all that is needed are the measurements of the angles, and then the remaining sides can again be calculated using trigonometry. And so on.

The advantage of triangulation is clear when you think about the alternative – namely, measuring the length of the meridian by laying a ruler or tape measure on the ground for the entire hundred-kilometre trip. This would be neither practical nor accurate over such uneven terrain; using triangulation, however, Maupertuis's party needed to make such a ground measurement only to find a baseline for one of the triangles. The chosen baseline was a fourteen-kilometre north-south stretch of the Tornio River, and the scientists waited until winter when the river froze, providing a flat surface that could be measured relatively accurately. Their 'tape measure' was a series of long wooden poles cut from the local fir trees, each pole being cut to exactly the same length.

When they had also measured the angles between the vertices of this reference triangle, they could use trigonometry to calculate the lengths of its other two sides. These two lengths could then be used as baselines for the adjacent triangles, and so on for the whole series of contiguous triangles. During the summertime journey up the valley, only the angles needed to be measured – the initial baseline measurement and all the subsequent calculations would be filled in later. The angles were measured by a large 'quadrant of two feet radius'; this was an instrument with a movable viewfinder attached to a quadrant (or quarter circle) on which degrees were marked on a scale.

Maupertuis recorded that while the baseline was being measured, 'the sun scarcely rose above the horizon, but the twilight, the white snow, and the Aurora Borealis supplied enough light for four or five hours of work daily'. During the autumn, and the particularly bitter winter of 1736–37, most of the team boarded with local people, and there seems to have been good will and respect on all sides, so much so that Émilie wrote a teasing note to Maupertuis, saying all his letters were full of praises for the 'Lapp women', and that 'apparently Clairaut has left me for one of them'. 'More seriously,' she added, 'tell me when you will be returning so we at Cirey can be worthy of you. We have become philosophers absolutely!'

On the way back to Paris in the late summer of 1737, Maupertuis's team was shipwrecked and many of their scientific instruments were damaged; fortunately, they managed to save their calculations (and it is interesting that Maupertuis also managed to save Émilie's letters). When they finally arrived home safely, Maupertuis was in his element – the returned hero who captivated Parisian society ladies with his exotic tales of adventure and his triumphant scientific news: when compared with La Condamine's earlier results, all the careful measurements made in the Arctic suggested that the earth was, indeed, oblate, and that the amount of flattening was closer to Newton's prediction than to Huygens's. Voltaire was typically direct and succinct: 'Maupertuis has flattened the Cassinis as well as the earth.'

Although Cassini's measurements and Huygens's calculations were clearly inadequate and were no longer in contention, it turned out that neither the theoretical predictions by Newton *nor* the physical measurements made by Maupertuis and La Condamine were completely accurate: today the accepted difference between the equatorial and polar diameters of the earth is around twenty-one kilometres or just over thirteen miles, compared with Newton's prediction of about seventeen miles (around twenty-seven kilometres) – that is, there is actually *less* flattening than Newton predicted, not more as Maupertuis had claimed. Clairaut was the one who ultimately discovered the reason for the discrepancy: Newton had assumed the 'fluid' in the earth to be homogeneous, although he had recognised it is not so in reality and had made some preliminary suggestions about how to make more realistic calculations. Clairaut extended these ideas, and his reworking brought the Newtonian prediction much closer to the modern measured value.

In the mid-twentieth century – through satellite observations – a second, much smaller bulge was discovered near the North Pole, so the earth is actually somewhat pear-shaped. Such an odd-shaped earth defies the intuitive logic of the centrifugal force argument, so it is not surprising that Newton did not consider applying his law of gravity to determine the size of this second bulge. But the precision of Newton's law itself is quite astounding, as I mentioned earlier. Even the assertion that all matter exudes a gravitational force has been experimentally verified, although for everyday objects, this force is so tiny it is negligible for most purposes. Nevertheless, in 1798 – more than a century after the *Principia* was first published – Lord Cavendish devised an extraordinarily delicate laboratory experiment that demonstrated and measured the gravitational force between two spherical balls of a given mass.

This is hindsight, however: despite the positive results of the Arctic and Peruvian expeditions, in the late 1730s the Continental battle for Newtonianism was not yet won.

## *Chapter 6*

# THE DANGER IN NEWTON: LIFE, LOVE AND POLITICS

In the first half of the eighteenth century the perceived 'trouble' with Newtonianism was manifold. There were not only the philosophical and technical issues that bothered Continental scientists, but also there were religious concerns over the idea that all matter is capable of exerting a gravitational force. (Recall that Cavendish did not verify this assertion experimentally until the end of the century.) In this context, I have already mentioned the *mathematical* gulf that separated Newton and Huygens when it came to calculating the shape of the earth via theorems that took account of the gravitational forces between all the particles of matter in the earth. But this aspect of gravitational theory also reveals the extent of the *conceptual* revolution wrought by Newton. Émilie and Voltaire's forthcoming *Elements of Newton's Philosophy* demystified some of these controversial ideas, but initially it seemed as if the book would not be published at all.

In his *Philosophical Letters* (which itself had not yet been published officially in France), Voltaire had contrasted Cartesian philosophy with that of both Locke and Newton. He noted that Descartes had believed 'thought' was equated with 'soul', and 'matter' was equated with 'extension' (because all physical objects exist in space as well as time). The first of these 'equations' was between supposedly self-evidently *immaterial* things, but Locke had challenged this notion by suggesting thought may be possible in 'purely material' beings without a soul. As for matter, in focusing on the size or 'extension' of objects, the Cartesians paid relatively little attention to material composition or 'mass', so the Newtonian idea that all material particles have mass, and therefore gravity, was philosophically and scientifically radical. But it was also theologically controversial, because Newton's claim that matter can attract other matter across empty space – with no evident agency other than the apparently inherent gravitational nature of mass – seemed akin to Locke's idea that perhaps matter without a soul can 'think'. To worried theologians, the ideas of both Newton and Locke suggested a philosophy of materialism in which there could be no immaterial afterlife for the soul.

As Émilie and Voltaire knew, Newton himself had argued against this kind of materialism, and Locke did not subscribe to it either. For instance, in the *Principia*, Newton spelled out the fact that he was only talking about the *mathematical* law of gravity – he was not speculating on its cause or metaphysical nature. But too few people read Newton closely enough, and Émilie realised that potential charges of materialism made the *Elements* 'a thousand times more dangerous than *The Virgin*' (Voltaire's notorious poem about Joan of Arc).

Émilie's anxiety over the *Elements* – and her fear that Voltaire would try to publish it without permission – had been particularly intense in January 1737, soon after they had finished the manuscript, because Voltaire was then in exile again. His dedication to Émilie at the beginning of the *Elements* had spoken of his joy in the dispassionate search for scientific truth, which seemed more appealing to him than the backbiting and hatred he endured as a playwright and poet. But by the end of 1736, two months after he had written his hopeful dedication, Paris had been buzzing with verses of another of his 'scandalous' poems, *Le Mondain*.

The title *Le Mondain* literally means 'The Socialite' or 'The Sophisticate', but the poem's meaning is perhaps closer to 'The Bon Vivant'. It paid tribute to the earthly pleasures of luxury, good food and wine, art and music – all the comforts of the modern 'Iron Age' of industry and technology, in contrast to the primitive, so-called 'Golden Age' in the Garden of Eden. Indeed, Voltaire's Adam and Eve were dirty, smelly and ill-nourished, without even a proper bed to sleep in. The intent of the poem was to redefine morality and happiness in terms of hard work and material abundance, as opposed to the prevailing religious idea that restraint and austerity in this earthly life were necessary for entry into the eternal afterlife. Consequently, Voltaire concluded his poem by saying that others may strive spiritually in the hope of entering Paradise, but he himself was already there. Of course, Cirey was his inspiration, as is evident from his frequent epistolary references to 'the earthly paradise of Cirey', and also from the testimony of many visitors. For example, historian Charles Hénault wrote, 'If one wanted to make a picture representing the pleasure of a delicious retreat – a sanctuary of peace, union, calmness of the soul, amenity, talent, reciprocity of esteem, the attractions of philosophy joined to the charms of poetry – then one would paint Cirey.'

Formal complaints about the supposedly blasphemous nature of *Le Mondain* had forced Voltaire to flee the country – Émilie rode with him to the closest border, from where he took the mail coach to Holland. It had been a tragic parting: Voltaire wrote of his dread of 'seeing the moment arrive when I have to leave her forever – she who has done everything for me, who has, for me, left Paris, all her friends and all the pleasures of her life, a person I worship and am bound to worship . . . the situation is horrible.' As for Émilie, she thought she would 'die of unhappiness'. But their loyal friend Argental, who had been a friend of Voltaire since his Jesuit school days, warned her that the only reason Voltaire had not been arrested long ago was the authorities' respect for the Du Châtelet family.

For instance, at the beginning of 1736 when *Alzire* was first performed at the Comédie Française theatre in Paris, many people had interpreted it as anti-Catholic, and there had been rumours and rumblings about Voltaire's supposed atheism. Recall that in *Alzire*, Voltaire had aimed to show that the true spirit of religion rests on 'natural' virtue – the natural, common-sense ethics whereby people do not kill or harm or steal from each other, but help each other where possible – rather than on prescriptive, institutionalised notions of 'religious' duty that are too often expressed through mindless obedience to ritual and dogma, Christian or otherwise. To flesh out this point, the play is set amidst the violent clash of cultures that occurred during the sixteenth-century Spanish conquest of Peru. In the play's denouement, Voltaire expresses his message of tolerance and natural virtue by having the fictional colonial governor, Gusman, discover on his deathbed that the true Christian 'regards all men as his brothers, treats them well and forgives their wrongs'. And so Gusman forgives the Peruvian warrior king, Zamore, who has dealt his mortal blow, asking forgiveness in return for all the cruelties he himself had perpetuated on Zamore's people.

Incidentally, *Alzire* is a five-act tragedy in rhyming alexandrine verse, where every line contains twelve syllables. Many of Voltaire's plays used this traditional poetic form, but some modern critics have considered it too stilted a medium for the powerful stories he was trying to tell. Perhaps this helps explain why his plays have not retained the timeless appeal of Shakespeare. It seems ironic that Voltaire himself had admired the 'force and fecundity' of Shakespeare's genius, and had thought his language was 'natural and sublime, without the least spark of good taste and without the least understanding of rules', as he put it in *Philosophical Letters*. At the beginning of the nineteenth century, before Romanticism took hold, formal 'rules' still dominated French theatre, so much so that on her first visit to Paris, Mary Somerville would comment on 'my admiration of Shakespeare, and my want of sympathy for the artificial style of French tragedy'. (However, Voltaire has recently made something of a literary comeback: there has been renewed interest in reclaiming him as a titan of French literature, and in giving his plays modern staging and interpretation. A contemporary stage-adaptation of his famous satirical philosophical novel, *Candide*, was performed around the world in 2009, the 250th anniversary of the novel's publication.)

At the end of 1736, with Argental's warning fresh in her mind, Émilie became alarmed when the text of *Alzire* was published, with a long dedication to 'Madame du Châtelet' that made it clear to everyone she was not merely Voltaire's disinterested patron. He obliquely acknowledged her assistance with *Alzire* – she had helped with the preliminary research, and she read and critiqued all its drafts – and he directly defended her against accusations that she had abandoned the responsibilities of her class and her sex in order to cultivate the sciences. He compared her with both Queen Caroline – the scholarly wife of George II of England, who had engaged in correspondence with Leibniz and Clarke (Newton's disciple) – and with the intellectual, sexually liberated Queen Kristina of Sweden, who had been Descartes's patron and a great supporter of the arts. But Émilie's pleasure in such a tribute was cut short, because Voltaire had also begun circulating his dedication from the *Elements*, which, she told Argental, 'is addressed to me under the name Émilie'. With both these dedications now in the public arena, and with Voltaire still in exile because of *Le Mondain*, Émilie decided, 'It may be necessary to abandon Cirey for a while, so they won't have an excuse to ask Monsieur du Châtelet not to give him refuge.'

By New Year's Eve, she had had no news of Voltaire for nearly two weeks, because his handwriting was distinctive and well known, and secrecy about his whereabouts was crucial to his safety. 'In the name of friendship, in my extreme unhappiness, calm me, reply to me, and have pity on me,' she pleaded to Argental. 'There is no-one but you who can truly instruct me [on how to help Voltaire] and whose opinion I value . . .' A few days later, she wrote again of her 'mortal fear' for Voltaire, and asked Argental to 'write him of the extreme danger he is courting, and of the wisdom of remaining incognito'.

She was particularly worried about the *Elements*: she begged Argental to advise Voltaire to leave out the 'materialist' chapter, and to tell him that if he insisted on publishing the book in Holland, then at least to send the manuscript to Paris at the same time, so that it did not appear as though he was trying to bypass the king's approbation. It was sound advice, but Voltaire felt she worried too much: 'He thinks I am afraid of my own shadow.' In fact, he refused to lie low even in exile: Émilie told Argental, 'They are playing *Alzire* in Brussels, Anvers and all the towns he has passed through . . . It is necessary to save him from himself [and] I employ more politics to lead him than the entire Vatican employs to control Christianity.

I count on you to second my opinions – he says all my letters are sermons.'

Eventually, Madame de Richelieu managed to obtain an official promise that the authorities would give advance warning if they decided to arrest Voltaire. And in the end, he returned safely from exile and no charges were laid: *Elements of Newton's Philosophy* had not yet been published, and the furore over *Le Mondain* had died down.

By January 1738 – more than a year after Émilie and Voltaire had finished their manuscript and several months after Maupertuis had returned from the Arctic – neither the *Elements* nor Maupertuis's account of his expedition had been published. Maupertuis's book was apparently held up by complaints from those loyal to Cassini and Descartes, in the wake of a quarrel between Cassini and Maupertuis's team that seems to have begun when Cassini cast doubt on the methods used in making the Arctic measurements. Maupertuis was extremely sensitive to criticism, and he responded with an excessively vicious attack on Cassini – so much so that the two men remained bitter enemies for the rest of Cassini's life. Meanwhile, Émilie wrote Maupertuis a consoling letter: 'I would have written to you sooner, Monsieur, if I had known you were unhappy, but you must feel superior to those not worthy of admiring you.' She firmly believed partisanship was blocking the spread of Newtonian ideas in France, adding, 'In the end, the French do not want Mr Newton to be correct – although it seems to me that, thanks to your efforts, part of his glory redounds on your country.'

She went on to speak of her fears that Voltaire – and perhaps even Maupertuis himself – would be arrested for their support of Newton, because Voltaire, too, was still having difficulties with officials, who had withheld permission to publish the *Elements* in France: '[They think] we are heretics of philosophy. I admire the temerity with which I say "we" . . .'

Émilie's help during the writing of the *Elements* certainly entitled her to say 'we', and, as I have already indicated, her ironical use of the word 'heretic' was also apposite: Voltaire's *Philosophical Letters* of 1734, and Maupertuis's semi-popular *Discourse* of 1732, had helped open the way for a more popular understanding of Newton's work, but many of their colleagues continued to reject it, throwing out the whole theory of gravity simply because it did not explain the cause of gravity itself, or because some of Newton's innovative applications of his theory were not yet fully worked out. Huygens, Leibniz, Fontenelle and Bernouilli never accepted the theory of gravitational 'attraction', but it was also resisted initially by some of the very men who would eventually help make it acceptable. For example, the great Swiss mathematician Leonhard Euler was extremely reluctant to accept any part of Newton's theory, although, like his friend Clairaut, he would ultimately do a great deal to develop and modernise Newton's mathematical language.

Euler was Émilie's age, and had studied in Basel under Bernouilli (Clairaut and Maupertuis's mathematical mentor). Bernouilli was probably the leading Continental mathematician of his generation: he was now in his seventies, but as a young man he had corresponded with Newton and Leibniz, and he was still a vigorous anti-Newtonian, pro-Leibnizian, neo-Cartesian. In fact, Newtonianism was so tenuous on the Continent that, in the 1740s, Clairaut himself would become temporarily convinced that the inverse-square law of gravity was wrong. After all, any new theory needs decades' worth of checking and testing before it can be considered to be 'true'. (By 'true', I mean the theory must offer descriptions of the world that give an excellent, predictive *approximation* to the reality we can see and measure with our senses and our technological instruments – but no theory can replace reality itself!) Besides, the *Principia* may have been bold, brilliant and extremely rigorous by the standards of the day, but it is true Newton's intuition was so profound that not even he could adequately prove all of his conjectures.

In fact, we now know that Newton had created a new approach to theoretical physics that was so innovative and fertile it would take more than a century for physicists to perfect and develop it, and half as long again for anyone to come up with a new theory that did not arise directly from Newton's work (namely, the theory of electromagnetism). Even without the benefit of hindsight, Voltaire and Émilie were aware that some of the early resistance to Newtonianism was neither scientific nor religious nor philosophical but simply nationalistic. In their *Elements*, they wrote perceptively of superficial critics who nitpicked Newton for not having got *everything* right, for not having explained *everything* about the universe: 'Is it self-love that makes them want to have the honour of writing against a great man? Wouldn't it be more flattering to be a disciple than an adversary? Is it because they are born in France that they are embarrassed to receive the truth from an Englishman? Such a sentiment is unworthy of a philosopher. For those who think, there are neither Frenchmen nor Englishmen: anyone who instructs us is our compatriot.' Algarotti also took up this theme in *Newtonianism for the Ladies*, having his narrator tell his readers, 'We will look impartially at philosophy . . . We will guard against illusion and prejudice . . . Happy is the society that mixes English good sense, French delicacy and Italian imagination.'

Of course, the French scientific establishment respected Newton's mathematical abilities and his experimental achievements on the nature of colour, and he had been made a foreign member of the Paris Academy of Sciences, just as Cartesians like Fontenelle and Mairan were members of London's Royal Society; in fact, as president of the Royal Society, Newton had personally sent Fontenelle a copy of his *Opticks*. Nevertheless, respect for ideas was one thing, but acceptance was quite another – and popular science writing played a role in changing this situation, and also in helping overcome philosophical and religious resistance.

As I indicated earlier, the most 'dangerous' part of Newton's theory of gravity was its supposed materialism, but Algarotti found that writing about Newton's experiments on light was similarly problematic. The Church was not opposed to experimentalism itself, but Algarotti made it clear he favoured empiricism over divine revelation. Consequently, he had had trouble getting permission to publish in Italy, and he had finally resorted to falsifying the publication details of *Newtonianism for the Ladies*, which appeared in 1737 in a clandestine edition. In subsequent authorised editions in the late 1740s and 1750s, he would progressively delete the most contentious material in the book, including some of its erotic allusions, and its implicit references to Lockean empiricism and to the link between Newtonian science and British political freedom (for example, he had stressed the need for freedom of speech and international cooperation in order to discover scientific truth). He even deleted 'Newtonianism' from the title, so that the 1750 editon was simply called *Dialogues on light, colours, and attraction*.

As for Voltaire's *Elements of Newton's Philosophy*, it had been printed in 1737, but by early 1738 the French royal censor had not replied to Voltaire's application for approval to publish. In March 1738, his publisher went ahead anyway and published the *Elements* in Holland, without Voltaire's permission. (Liberal Holland had long served as a place where writers could publish in any language works that had been proscribed in their own countries; for instance, after the Inquisition had banned him from expressing scientific opinions, Galileo had taken the Dutch option to publish his groundbreaking work on falling motion, *Two New Sciences*.) The unauthorised edition of the *Elements*, with the populist title 'Elements of Newton's philosophy for everybody', did not please Émilie and Voltaire – it was not their final version, and the last five chapters were either missing or had been rewritten by someone else. On the positive side, the 'blasphemous' section on the supposed materialism of Newtonian philosophy was one of the missing chapters, and so the French chancellor accepted that Voltaire had not deliberately published the book in Amsterdam in order to sidestep official permission from the French king.

Soon Voltaire succeeded in publishing a corrected edition in France, but his initial difficulties, together with the earlier drama of *Philosophical Letters*, upset him so much that five years later he was still bitter about what he and other writers suffered. For example, in 1743 he wrote to his friend Argenson bewailing the persecution of a seventy-year-old historian, abbé Lenglet de Fresnoy, imprisoned in the Bastille for his view of history. (Lenglet was imprisoned several times, partly because of his caustic style and his provocative reaction to royal censorship.) Moving on from Lenglet, Voltaire referred indirectly to his own problems: 'What height of barbarity, and what excess of petty meanness, not to permit the publication of books explaining Newton, or that say the daydreams of Descartes *are* daydreams.' Several months later, he told Martin Ffolkes (who was president of London's Royal Society) that he had been a 'martyr' for Newton. Voltaire had met Ffolkes during his exile in England sixteen years earlier, and he continued his letter in rusty English: '[In France] I could never obtain the privilege of saying in print that . . . vortices cannot be intirely reconcil'd with mathematics [or] that there is an evident attraction between the heavenly bodies . . . But the liberty of the press was fully granted to all the witty gentlemen who teach'd us that [gravitational] attraction is a chimera, and vortices are demonstrated.'

From a modern point of view, there is much that is confused or simply wrong in *Elements of Newton's Philosophy*, but overall, it remains a creditable, relatively accurate exposition of early eighteenth-century science. When the authorised, corrected edition was published, Émilie sent a copy to Richelieu, saying, 'I know one can make many criticisms of this book, but there is nothing better in French on these matters.' She also told Maupertuis she knew it did not deal with the most profound details of Newton's work, but that the title, 'Elements', said it all: the book was directed not at specialist scholars but at Voltaire's peers in the Republic of Letters, the 'people who think'. She would later publish an anonymous review of the book in the prestigious *Journal des savants*, in which she praised the *Elements* for its lack of complex algebra, which helped demystify Newton's gravitational theory. But she also noted Voltaire's extreme form of Newtonianism (she and Voltaire accepted each other's critiques as an important part of intellectual life: for example, she herself disagreed with some of Newton's ideas on light, as I will show in the following chapters).

As for *Newtonianism for the Ladies*, Émilie thought Algarotti had 'overburdened the truth with fripperies'. She wrote Richelieu: 'The *Dialogues* of Algarotti are full of *esprit* and knowledge – he wrote part of it here [at Cirey] – but I confess to you that I don't like its style on scientific matters: the love of a Lover that decreases in the ratio of the square of the time [away from his beloved, and also in proportion to] the cube of the distance between them, appears to me to be difficult to stomach.' This apparently refers to Kepler's third law, relating the time of a planet's yearly orbit to its average distance from the sun – recall the *Elements* had explained this law by showing how to calculate the distance between Venus and the sun. No wonder Émilie disliked Algarotti's fatuous, coquettish metaphor! But she was not Algarotti's intended reader: many of her lesser-educated sisters no doubt enjoyed his entertaining overview of the history and nature of natural philosophy. However, perhaps the majority of women read Algarotti's book simply to be fashionable: among many of the ladies of Milan, for example, *Newtonianism for the Ladies* inspired a craze not for Newtonian science but for English tea and hats. It also fostered an interest in learning English, languages being considered suitable pursuits for women.

Émilie was so angry at Algarotti's trivialising approach she wrote in disgust to Maupertuis, too, bemoaning Algarotti's argument that 'a lover who has not seen his mistress for eight days loves her sixty-four times less! I don't know what pleasantry he would have come up with for the inverse-square law of distances [in the theory of gravity].' But she continued generously, 'All in all, it is the work of a man of much *esprit* and who is master of his material.' Consequently, she was extremely hurt that Algarotti had reneged on an apparent promise to dedicate his book to her, choosing Fontenelle instead. Of course, Algarotti had poached Fontenelle's device of the philosopher-narrator explaining science to a curious 'marquise', so a dedication to him was appropriate. But Émilie made the astute point that Algarotti's dedication to Fontenelle was 'singular' in a book called *Newtonianism for the Ladies*, because 'he is neither a woman nor a Newtonian'.

The style of Algarotti's book *was* rather demeaning to the real-life marquise du Châtelet, although presumably he did not set out to do this deliberately but was simply too insensitive to realise that his flighty heroine fitted perfectly into the public perception of Émilie as an intellectual upstart who, just like Algarotti's marquise, could never enter fully into the mathematical inner sanctum of physics.

The book's sexist approach may seem surprising given Algarotti's close friendship with Émilie and also his respect for another female scientist, whom he mentioned obliquely in the first chapter of his book: in setting the scene for the first 'dialogue', he used the device of arousing his marquise's scientific curiosity by having his narrator read her a poem about light and colours – a poem the narrator has written 'for the glory of our Bolognese savante'. Algarotti had written the poem some years earlier, to celebrate the graduation of the young Italian Newtonian, Laura Bassi, who had received a degree in philosophy at Bologna in 1732, when she was twenty-one years old. She was only the second woman to gain a modern university degree, after Elena Piscopia, who took a philosophy degree in Padua in 1678 – but there would be few other female university graduates in the world until the eve of the twentieth century.

Several years younger than Émilie, Bassi was a prodigy who had been given an excellent education by her father. In the 1730s, when Algarotti was writing his book, Bassi was lecturing at the University of Bologna in philosophy, including 'natural philosophy' or physics. But her role was essentially a ceremonial one, aimed at enhancing the glory of her city – she was called the Minerva of Bologna, and she gave public rather than academic lectures. Algarotti no doubt discussed her at Cirey, presumably prompting Voltaire to refer to Émilie as 'the Minerva of France'.

At the age of twenty-six, Bassi married physicist Giuseppe Veratti, and the town fathers felt her symbolic virginal 'Minerva' status was somewhat tarnished; the university now expected that she spend time at home rather than in the lecture theatre, caring for her family – she would have eight children, of whom five survived childhood. Undaunted – and encouraged by her husband (with some university funding) – she set up a laboratory at her home, where she gave lessons in experimental science. Her laboratory attracted not only students but also experienced physicists, including international visitors. Not surprisingly, Laura Bassi did not like the frivolous style of *Newtonianism for the Ladies*.

Although the road to publication of *Elements of Newton's Philosophy* had been frustrated by censorship, the book would prove extremely influential in popularising the theory of gravity on the Continent. 'All Paris studies and learns Newton', wrote an enthusiastic reviewer. By 1743, the book's success would facilitate Voltaire's election as a fellow of the Royal Society of London, and ultimately to a score of other European academies.

Needless to say, no such honours came to Émilie: apart from the fact that she was not formally named as co-author, few scientific academies would admit women until the mid-twentieth century. (The Royal Society admitted its first woman in 1945, and the Paris Academy of Sciences in 1962.) But Émilie was not deterred. She and Voltaire had prodigious intellectual energy, and while they were waiting around for the publication of the *Elements*, they had been working on another project – a project, it turned out, that would yield Émilie's first taste of fame.

*Chapter 7*

# THE NATURE OF LIGHT: ÉMILIE TAKES ON NEWTON

Émilie and Voltaire's new project was an entry for the French Academy of Sciences' 1738 essay competition for the best paper on 'the nature and propagation of fire'. This topic included the nature of light, because 'fire' was regarded then as a generic term for any source of light and heat; consequently, preparation for their essays overlapped with their work on the *Elements*, the first half of which had been devoted to theories about the nature of light. In their essays, Émilie and Voltaire would also address the specific nature of fire itself, and even such things as the nature of ice, which they regarded as the opposite of fire; but their ideas in this context are now so outdated that I will focus only on their analyses of heat and light.

As part of her 'Newtonising at Cirey', Émilie had already repeated many of the experiments Newton had meticulously described in *Opticks*. Firstly, to see the spectrum, he had produced a beam of white light by letting sunlight shine through a hole in his closed window shutter, in such a way that it passed through a triangular glass prism set up on a table near the window. When the light from the prism reached a wall several yards away, the magical rainbow-coloured band appeared. It is a common enough image today, although none the less beautiful, but initially it was so unexpected that its discovery was surprisingly slow. After all, even Descartes – who was more of a rationalist philosopher than an experimenter – had thought of shining a light through a prism, but he projected it onto a piece of paper only two inches (about five centimetres) away, and all he saw were a red and a blue spot. Following Descartes, Robert Hooke shone a beam of light through a beaker of water and projected it onto a sheet of paper two feet (or 60 centimetres) away, while Robert Boyle shone prismatic light onto the floor, about four feet away. They saw only a circle of light with coloured fringes.

The colours obtained in these ways were regarded simply as impurities or artefacts of the prism, but Newton had begun to realise that colour was inherent in white light itself. He noticed that when ordinary white light was shone onto thin gold leaf, some of the light was reflected and scattered by the leaf back to his eyes in such a way that the leaf appeared to be gold (or yellow), but that if he viewed the leaf from the other side – that is, if he looked at the back of the leaf – then it looked blue, as if the white light passing *through* the leaf were missing its reflected yellow component. To test his idea, he worked out how to obtain a spectrum – by projecting light from a prism onto a wall 22 feet (or about seven metres) away.

It can be difficult to count the different colours in the spectrum, as is apparent when looking at a rainbow, which is created because the droplets of water in the air act like tiny prisms. Sometimes, Newton thought he could identify six separate colours, sometimes he thought there were seven. Ultimately, he was drawn to the number seven for a metaphysical reason rather than a purely experimental one – namely, because a spectrum of seven colours was analogous to a musical octave, and he rather liked the ancient Greek idea of a harmonious universe whose structure reflected musical harmony. This is an example of the way he reverted to the old-style use of untested hypotheses in his tentative speculations on light, in contrast to his rigorous methodology in the *Principia* – and for three centuries Newton's list of seven colours was accepted. Today, however, physicists generally accept six colours, leaving indigo out of the list. (Actually, there are six 'families' of colour rather than six single colours, because in a spectrum there is a continuous shading of colour from red through to violet.)

Émilie and Voltaire had equipped their Cirey laboratory with prisms and lenses, and they both took delight in verifying various experiments described in *Opticks*. In particular, they repeated the landmark experiment in which Newton passed each colour of his spectrum, one at a time, through a pinhole in a sheet of black cardboard. He did this in order to create a series of separate beams, each of a single colour, that could then be passed through another prism, in order to prove none of the spectral colours splits into yet more colours.

This crucial experiment had also been repeated by Algarotti in 1728, in answer to Italian critics like Giovanni Rizzetti, who did not accept Newton's results because others had tried to replicate them and failed: on passing through a second prism, the colour red, for example, did not remain red as Newton had claimed. Optics had been Algarotti's specialty when he was a university student in Bologna, and he realised the problem lay in flaws in the prisms used – the Italian experimenters had used Venetian glass, which contained certain impurities – and he obtained prisms of British glass in order to successfully complete his experiments. Which, not surprisingly, led to the ironical claim that 'Newtonianism only worked with British prisms', a claim that also raised legitimate questions about how to decide whether or not experiments are objective.

Even in the late 1730s, doubts lingered, much to Émilie's surprise. In December 1738, she would tell Maupertuis that she thought Newton's experiments on refrangibility and on mixing colours were so careful that surely there could be no doubt about them. She said she had replicated some of them herself, but she mentioned the opposing results of her compatriot Du Fey, adding she doubted he had a case. Nevertheless, she was willing to suspend her judgment on Du Fey until she saw his experiments for herself. Of course, there have now been so many independent replications of Newton's spectrum experiments that they are universally accepted as proven facts. But in the face of Rizzetti's criticism back in 1722, the then seventy-nine year old Newton had felt compelled to arrange for a colleague to demonstrate these experiments in public. This was a full fifty years after he had first presented his results to London's Royal Society!

But Newton had endured criticism from the very beginning. This was partly because his unexpected discovery – that it was colour, not 'white' light, that is fundamental in nature – was a shock to the established view. Huygens, for example, took the same approach he later used against the *Principia*, insisting that Newton's spectrum of colours had no mechanical explanation, and therefore proved nothing about the true nature of light. Eventually, he appeared reluctantly to change his mind, accepting that Newton had, indeed, discovered a new phenomenon, even if he had not explained its underlying cause.

Newton was also criticised for more legitimate reasons – namely, that his attempts to explain his results were not always successful. For instance, the colour of an object is not always a simple matter of reflection and transmission: not all objects behave like gold leaf, as contemporaries like Boyle and Hooke realised. The colour of an object depends on several things, including the colour of the incident light, the temperature of the object, and the surface of the object. Nevertheless, Newton was right in suggesting that colour resides in white light itself, and that an object's colour arises from those parts of the illuminating light source that are not absorbed by the object but are scattered, transmitted or reflected by it. This is also evident when coloured rather than white light is shone on the object: for instance, if, in a darkened room, a beam of red light is shone on a blue object – that is, an object that absorbs all colours but blue – there is no blue light in the illuminating beam, so there is no blue light for the object to scatter or reflect and hence it will appear black.

When he was still in his twenties, Newton had applied one of his key discoveries on colour to invent the reflecting telescope, which was an improvement on the earlier refracting telescopes. (Actually, James Gregory had described a design for such a telescope several years earlier, but he had not been able to turn his idea into a working model as Newton did with his own simpler design. It was also Newton who explained the physics behind the distortions produced by refracting telescopes, as will be seen shortly.) The reflecting telescope used the reflective properties of a concave mirror to collect and focus light from a star or the moon. The law of reflection is that if an oblique light ray hits a surface at a given angle, it will be reflected away at an angle of the same magnitude (but in a different direction, as illustrated in Figure 2 in the Appendix). A curved mirror makes use of this property, as you can see in Figure 3: when the different parts of a horizontal light beam meet the mirror at different angles and are thus reflected at different angles, the curve of the mirror ensures all the reflected rays meet at the point of focus.

The simple mathematical law of reflection had been known for at least two thousand years: Euclid had mentioned it, and a geometrical analysis of reflection from a spherical mirror had been discussed by tenth-century mathematician ibn-al-Haitham. Al-Haitham had also made pioneering studies in refraction, but the modern mathematical law of refraction in various media was discovered experimentally in 1621, by Willebrod Snell. Even before this mathematical law was known, experimenters were skilled in using refraction in the manufacture of spectacle lenses, and it was a spectacle-maker, Hans Lippershey, who is credited as the first commercial manufacturer of telescopes, beginning around 1608: these early telescopes – and the ones invented by Kepler and Galileo around this time – had used refracting lenses rather than reflecting mirrors. But refracting telescopes suffered from a coloured distortion caused by refraction – a distortion that Newton realised would not occur with the simple reflection of light from a mirror.

Recall that refraction is the bending of light when it passes from one medium to another, as you can see when you dip a pencil or a straw into a glass of water: if you bend your head so as to observe the pencil at an oblique angle, it appears disjointed at the water's surface. (This is illustrated schematically in Figures 4 and 5 in the Appendix; the focusing of light by a refracting lens is illustrated in Figure 6.) Snell (and also Descartes, who had published a geometrical analysis of optics) had known that the amount of bending of an ordinary (white) light ray during refraction from air into a denser medium like glass or water is different for each different medium, but Newton showed that in addition, the different *component* colours of light are each bent by slightly different amounts in the *same* medium. This, of course, is why a beam of ordinary light can be split into colours when it is refracted at the glass surface of a triangular prism – but Newton realised this had ramifications when it came to using a glass lens: the fact that each component colour of an incoming light beam is bent slightly differently means the point of focus for each colour must be slightly different from all the rest, so a single lens cannot produce a sharp image, but only an image blurred by a coloured fringe. In a reflecting telescope, by contrast, the light is focused by pure reflection – and reflection does not separate the colours of a beam of white light.

Newton's reflecting telescope prototype was barely six inches (or fifteen centimetres) long and its mirror was about one inch (or twenty-five millimetres) in diameter, but it gave a magnification of about thirty-five times, which was as good as that of a refracting telescope six feet long (or almost two metres). Today, astronomers use reflecting telescopes with huge mirrors whose diameters can be eight metres or more (and to create even bigger telescopes, *arrays* of mirrors are used). These telescopes are so powerful they could detect a candle flame on the moon, so they can see light from stars that are almost unimaginably distant. Refracting telescopes have also improved: in fact, not long after Newton's death, the 'achromatic' lens was invented to reduce the distortion due to refraction.

Newton's invention led to his election to London's prestigious Royal Society, but his research paper explaining his discoveries on colour – including those that underlay his telescope – fared less well. This was not only because of his novel ideas on colour, and because his attempts to explain his experiments were not always correct; it was also because he favoured a particle hypothesis to explain the nature of light. In the seventeenth and eighteenth centuries, no-one knew what light was made of – Descartes and Newton thought it was made of luminous particles, while Hooke and Huygens suggested it was some sort of vibration in the mysterious cosmic ether. Newton favoured the particle idea because it seemed the most logical way to explain the fact that light produces sharply defined shadows – it does not seem to bend into the shadow as water and sound waves do. You can see this type of wave bending when streamlines in a river flow into the 'shadow' space behind a solid obstacle like a rock. Particles of light, on the other hand, would either be absorbed by the obstacle or they would bounce off as reflected light, so there would be no 'leakage' into the shadow.

The physics of ordinary particles also seemed to explain the law of reflection itself. Recall that this law says that the angle made by an oblique beam of incoming light as it strikes a reflecting surface has the same magnitude as the angle made by the reflected beam as it leaves the surface. This can easily be explained in terms of an elastic collision of particles, as exemplified by a billiard ball that reflects obliquely off the raised side of a billiard table. However, Huygens had given a beautiful geometrical model of how light 'waves' might produce reflection *and* refraction. Nevertheless, he did not have a general theory that explained a wider range of properties of light, so in the seventeenth and eighteenth centuries, opinion was divided on the merits of the two models: waves or particles.

Newton was so upset at being criticised for his paper that he wavered between vowing never to publish anything again, and wanting to write a book that would establish his optical discoveries once and for all. He decided on the latter course, but while he was preparing his manuscript for publication, a lighted candle fell over and burned all his papers. He abandoned the project and swore that he would all but abandon science itself: the whole business of constantly having to defend his experiments led him to conclude that he had become 'a slave to Philosophy', and he wished to 'resolutely bid adew [sic] to it eternally, excepting what I do for my own private satisfaction . . .'

Years later, it was only because his friend Edmond Halley insisted he publish his extraordinary insights on planetary motion that Newton, then in his forties, embarked on the writing of the *Principia*, whose publication Halley paid for because the Royal Society had no spare funds. Halley actually made about ten pounds profit on the first edition, which ran to no more than about five-hundred copies!

Emboldened by the rapturous reception his *Principia* received in Britain, Newton eventually wrote up another account of his work on light, published as *Opticks*, the book Émilie and Voltaire were now using in their own laboratory. They also owned a reflecting telescope: the rooftop at Cirey provided a good vantage point, and the dark night sky gave them a clear view of the moon and the stars.

Voltaire's ongoing business investments (beginning with his lottery winnings and augmented by the money he made as a bestselling author) had made him very wealthy, and he had spent the equivalent of more than a million modern American dollars on experimental equipment, making the laboratory at Cirey one of the best in Europe. 'We are in a century where one cannot be learned without money,' he told Moussinet, the Paris-based friend who helped him with his business affairs, and who purchased the scientific instruments he ordered and then sent them out to Cirey. By contrast, Newton had made most of his equipment himself, even his telescope: in *Opticks*, he spent two pages describing his painstaking method of grinding and polishing a piece of metal in order to form a mirror of a specific curvature that would appropriately focus light from a distant star.

Although Voltaire had spared no expense in equipping Cirey's laboratory, he spent yet more money specifically for the essay competition. His approach to the topic of light and heat was firmly based on Newton's idea of luminous particles (or 'corpuscles'): he was so in awe of Newton that he believed in the reality of these particles, even though Newton never did develop his model in any detail. This was partly because he knew it had limitations, as Hooke and others had pointed out in their criticism of his first paper on light. In *Opticks*, he focused on experiment rather than theory, relegating most of his theoretical ideas to an appendix of 'Queries', in acknowledgment that these ideas were purely speculative.

The key point for Voltaire was that Newton had assumed these tiny luminous particles were material, so they had mass, which meant they had weight, albeit an infinitesimal amount. ('Weight' is the gravitational force on a material body, and according to Newton's law of gravity, there can be no weight without mass. Conversely – as space travel has shown and as Newton's law predicted – there can be no weight without gravity: weight is defined as mass times gravitational acceleration.) Therefore, since heat and light were both aspects of 'fire', Voltaire wanted to prove that heat, too, had weight. He was inspired by the experiments of Petrus van Musschenbroek in this regard – Voltaire had met Musschenbroek, and his equally Newtonian compatriot Willem 's-Gravesande, in Holland during the *Le Mondain* scandal – and he set about replicating one of Musschenbroek's experiments, which suggested that heated metals gained weight as well as heat.

With Émilie's help, Voltaire used one of Cirey's huge forges to heat various metals, which he weighed before and after heating to see if he could confirm Musschenbroek's finding. He had bought a variety of state-of-the art weighing scales, because he experimented on amounts of iron ranging from an ounce (twenty-eight grams) to two thousand pounds (almost a tonne). With such a huge investment in experimental physics, he was dearly hoping to win the essay prize – a dream that was not as far-fetched as it may sound: until the late nineteenth century, it was not uncommon for amateurs to make scientific discoveries. In fact, early modern science was often considered a (wealthy) gentleman's pastime rather than a profession.

Another of Newton's speculative ideas was that refraction might be caused by a force of attraction between the particles of light and the medium of refraction. Newton conceived this idea because when light passes through glass (or water) out into a more rarefied medium like air, it is bent towards the glass, as though it is being pulled back into the denser medium. (You can see this in Figure 4 in the Appendix, where, if you follow the light ray up out of the water and into the air, it is bent towards the water.) In the *Elements*, where he explained the theory that informed his experiments on fire, Voltaire tried to visualise this idea by showing that if a lead ball is thrown obliquely downwards through the air above a body of water, then when it enters the water it is slowed down and buoyed up. But once the sideways force of the ball's original motion is spent, it will fall straight downwards, under the influence of the earth's gravity. The 'effective' path of the ball – from its starting point to its endpoint when it has fallen to the bottom – mimics the bending in the path of a refracted light ray (as you can see in Figure 7 in the Appendix). Voltaire applied the same analogy to light corpuscles, suggesting the mass of the water itself – which is much greater than that of air – gravitationally pulled the refracted light particles towards it.

He then concluded that, once inside the water, the particle would have a greater speed than it did in air, because of the gravitational force of the water. It was a valiant attempt to give substance to Newton's tentative hypothesis that some kind of force of 'attraction' was responsible for refraction, and that it caused light to speed up in the refracting medium. But Voltaire seems to have taken Newton's 'Queries' too literally. For instance, it is true that in the *Principia*, Newton had mathematically proved that bodies moving under a centripetal force will move faster in a denser medium, and he expressly stated that this theorem could be applied to the paths of light rays in a refracting medium. But he was careful to add that he was 'not arguing at all about the nature of the rays (that is, whether they are [material] bodies or not), but [was] only determining the trajectories of bodies that are very similar to the trajectories of rays.' In other words, all Newton had claimed was that *if* refraction were due to some kind of attractive force, then the speed of light would be greater in the denser medium.

This result conflicted with the wave theory of Huygens, which predicted light travelled more *slowly* in denser media. But the jury was still out on the matter, because at that stage, no-one had been able to measure the speed of light in different media. Voltaire was therefore entitled to use Newton's ideas, even if he did seem to take them more literally than did Newton himself. And he was not alone: over the next few years, both Clairaut and Maupertuis would publish mathematical papers on the application of Newtonian attraction to refraction. Émilie would summarise Clairaut's work in her commentary on the *Principia*, but back when she and Voltaire were trying to measure the weight gain of heated metals, she doubted fire and light were made of material particles at all.

Émilie had not intended to enter the essay competition herself, being content to help Voltaire as usual. But she felt he was on the wrong track in following Newton's and Musschenbroek's supposition that light and fire had 'weight' or 'heaviness', because Voltaire's own experiments did not show any consistent result. Some metals seemed to gain weight on being heated, but others did not – and no matter how much iron Voltaire melted, no matter how many sets of scales he purchased in order to weigh his metals ever more accurately, he was not able to arrive at a definitive conclusion. Émilie realised that other researchers, too – including Boerhaave – had failed to confirm Musschenbroek's experiment. Besides, even if metal did gain weight on heating, Émilie believed it was impossible to know whether this extra weight was due solely to 'particles' of heat as Voltaire assumed, or whether extraneous particles of wood or carbon had contaminated the melted iron, thus adding to its measured weight.

Voltaire was so passionately Newtonian he would not listen to her, and so – with only one month to go until the deadline – Émilie decided to write her own paper. It was an extraordinary thing to do – to enter a competition that, as far as anyone knew, no woman had entered before. She was motivated not so much by ambition as by the need for truth – and by the realisation that her lover and mentor had lost his way, and that not even Newton could help him. She was so worried about hurting Voltaire's feelings by disagreeing with him that she decided to write her paper in secret, which meant that often she had to help Voltaire during the day and then stay up most of the night to do her own work. The secrecy of her project also meant she could not use the laboratory or the forge to do any experiments of her own, apart from some simple ones she carried out in her own room.

One of the most original parts of Émilie's essay was her suggestion that heat is related to colour – that different coloured objects absorb heat at different rates, other things being equal. She related this to a second, more specific suggestion: that the different colours of light itself have different temperatures. She thought of the first of these conjectures because of the obvious difference between black and white objects.

Recall Newton's discovery that 'white' light, like that from the sun, is made up of all the colours of the spectrum, and that an object's colour depends on the way it responds to these different components of the illuminating light beam. The colour of an opaque blue object, for example, is the result of blue light being scattered or bounced away by the object, the other colours all being absorbed by it. This blue light reaches our eyes and so we perceive the object as blue. Green objects appear to be green because they return mostly green light back to our eyes; red objects return mostly red light, and so on. As for white items, they return white light to our eyes – that is, all the colours of the spectrum. This power of reflection evidently keeps white objects cooler than black ones, which absorb most of the light that falls on them: such objects look black because they are reflecting very few rays back to our eyes, but are absorbing them instead. So Émilie wondered whether or not it was possible to detect different degrees of heating among coloured objects, not just black and white ones.

It was a cutting-edge idea: no-one had yet done an experiment that would link colour directly with heat. Without recourse to proper experimental equipment, she tested her hypothesis by dyeing different parts of a bed sheet with what she called 'the seven colours of the prism'. Then she wet the sheet with water, and let it dry in front of the fire, keeping it stoked so as to give the same heat throughout the experiment. She timed how long it took each different coloured piece of sheet to dry – and she found violet dried the fastest, followed by all the colours in the order of the rainbow, with red taking the longest to dry.

It hardly seems like a rigorous experiment, but as you may have noticed when doing the washing, it is surprisingly easy to apply Émilie's experiment to black and white cloth of the same type – to wash them and let them dry in the sun, and to show by simple touch that the black cloth dries much more quickly. Of course, it is harder to differentiate the drying rates of the different coloured cloth, because the differences are less extreme than between black and white, but nevertheless, Émilie had come up with a workable experimental design.

Today we would explain her result in terms of energy. Molecules in material substances (and especially in liquids and gases) move and vibrate, so they have kinetic energy. The term 'kinetic' refers to motion: kinetic energy is the energy an object has because it is moving, and it is related to the speed of the object. (Its mathematical definition is $KE = \frac{1}{2} mv^2$, where $m$ is the mass of the particle, $v$ is its velocity, and $v^2 = v \times v$.) Evaporation begins when the kinetic energy of some of the water molecules becomes strong enough for them to overcome the forces of electrical attraction that bind the molecules together in liquid form. In liquids and gases, the speed of molecules is related to temperature, so the hotter the water, the faster it evaporates, which suggests Émilie's experiment showed that the fast-drying violet-coloured cloth absorbed heat more quickly than the other colours, with the red cloth absorbing heat at the slowest rate.

However, the idea that heat is a form of energy had not been formulated at that time – nor had the electrostatic atomic model of molecules – and so Émilie linked the process of evaporation of water with reflection: 'The water, by the same fire [or amount of heating], was drawn from the pores of these colours in this order, beginning with those that dry the most quickly: violet, indigo, blue, green, yellow, orange and red. The reflection of rays follows the same order, and that cannot be otherwise, because the body that absorbs the least number of rays is surely the one that reflects the most.'

The cryptic, rather confused nature of this conclusion is no doubt due to the haste with which Émilie composed her essay, and on its own, this statement does not seem to make much sense. It is true that white objects absorb the least heat because they reflect the most light. But white objects take the *longest* to dry, whereas Émilie seemed to make a link between the (apparent) fact that violet light was reflected 'most' and the fact that the violet cloth dried fastest. As for the 'order' of reflection, she may have been confused by Newton's experiment on total internal reflection inside a prism, in which violet is reflected 'first' because it is bent the most on refraction: internal reflection occurs when a light ray coming through the glass approaches the glass-air surface at such an angle that it does not pass out into the air but is reflected back into the prism.

Nevertheless, she had an intuitive grasp of the correct analysis, because to further test her conjecture that heat and colour are related, she proposed another experiment. The idea occurred to her by analogy with our physical senses: if different parts of the spectrum appear visually in the form of different colours, then why wouldn't they *feel* different, too? Because red seems to be the most 'intense' colour visually, she suspected it might also feel the warmest, and that violet would feel the coolest. This conclusion actually follows from her sheet experiment. To heat up more quickly, Émilie's violet cloth must have absorbed the warmest colours of light, scattering away the coolest light, namely violet. Conversely, the slower-warming red cloth must scatter away the warm red light while absorbing the cooler rays.

Émilie did not make this connection; instead, she suggested a more direct experiment – 'very curious, if it is possible' – in which one would 'gather separately enough rays of each homogeneous colour to prove whether the fundamental rays that excite in us different sensations of colour would also have different burning properties – if the reds, for example, would give a greater heat than the violets, which is what I suspect'.

If only she had been able to trust Voltaire – if only he had kept his ego in check and listened to her ideas – she could have asked him to help her carry out such an experiment, which, if successful, would have made her famous for an important new discovery. As it was, she had no recourse in her essay but merely to suggest that others might be able to do this – perhaps even 'the philosophers who must judge this essay'. Apparently none of the Academicians listened to her advice, because such an experiment would not be carried out for many decades, after which time a number of experimenters – including Alexis-Marie Rochon in France and William Herschel in Britain – independently came to the same conclusion as Émilie had done and confirmed it experimentally.

Herschel was a colleague of Mary Somerville, who discussed his discovery in her book *On the Connexion of the Physical Sciences*. For the moment though, it is enough to note that Émilie's insight on the relationship between colour and heat shows the quality of her developing scientific imagination. Indeed, when it came to the question of light, fire and weight, the fact that she was prepared to disagree with both Voltaire and Newton showed just what an independent thinker she had become. She flatly rejected the conclusion Voltaire expressed in his essay and in the *Elements*, that 'Fire is matter; therefore it has weight, and light is nothing but fire'. In her essay, she said that 'fire has no weight, or if it does, it can never be perceived by us.'

She had reached this conclusion not only because of the inconclusive experiments of Voltaire and others on the weight of heated bodies, but also because of another experiment she managed to do in secret, in her own room. It was a simple illustration of the tendency of fire to rise – quite the opposite, she said, of tending towards the centre of the earth as ordinary objects do. 'If you put a plate on one of the large [open] glass cylinders that serve in summertime to cover candles, and you leave a lighted candle under the cover,' she wrote, 'then . . . the flame conserves its conical shape until the moment it is extinguished.' She went on to explain, 'This is because the candle contains enough fire to oppose [any] natural [gravitationally induced] tendency of the flame towards the centre of the earth.' In other words, fire had no (net) weight, but it had what she called its own 'superiority of force' – we would now call it energy – which gives it 'lightness' rather than heaviness.

Émilie also commented on the tendency of heat to rarefy objects (to make liquids evaporate into gas and to make gases expand, for example), so that 'rarefaction by fire is one of the laws of nature, without which everything in nature would be compact [because it would be subject to gravity only]'. A more modern example of this idea is related to 'dead' stars that have used up all their nuclear fuel, the burning of which gives both light and size to the star: without the counterbalancing energy of these nuclear reactions, the stars collapse under their own weight to form either black holes or compact 'neutron stars'. Émilie was far from predicting such things, but for all the naive simplicity of her language, it is clear she was struggling to formulate a concept of energy – a concept that would not be fully developed for another hundred and fifty years.

Despite Émilie and Voltaire's months of hard work conducting original experiments and critical analyses for their essays, the prize was shared by three Cartesians, one of whom was Euler. Émilie wrote to Maupertuis with characteristic emotion, 'We are in despair at the judgment of the Academy of sciences; it is hard knowing that the prize was shared and M. de Voltaire did not have a share of the cake . . . Euler is a Leibnizian, and consequently a Cartesian: Voltaire is angry that the spirit of partisanship has such credit in France.' He had some grounds for his suspicion, because Euler's was the only one of the three prize-winning papers that contained anything significant.

Entries for the essay competition had been submitted anonymously, but several weeks after the winners were announced, Émilie's secret was out. She wrote Maupertuis, 'I believe you were astonished that I had the boldness to compose a memoir for the Academy. I wanted to try my hand incognito – Monsieur du Châtelet was the only one whom I took into my confidence, and he kept the secret so well that he told you nothing of it [when he saw you] in Paris.' She said that helping Voltaire had inspired her to want to try a scientific career herself, but that she had not told him about her essay because she thought it might displease him that she was 'combating all his ideas in my work. I only told him when I saw in the gazette that neither he nor I had a share of the prize. It appears to me that a rejection shared with him is very honourable.'

Despite her worries about upsetting him, Voltaire's reaction to Émilie's deception seemed to be one of pride in her achievement. He knew that Maupertuis had been impressed with his own essay, but that the other Academicians did not agree because of his Newtonian approach. But he was sincerely impressed with Émilie's work, and he wrote to a friend, 'Without her excessively bold opinion that fire is not material, [she] would have deserved the prize.' He even went so far as to suggest to his friends in the Academy that her essay (along with his own) would be worth publishing. The Academy did occasionally publish the papers of 'runners-up', and in this case, it eventually decided that, although it did not agree with the ideas in these two essays, they each showed such a wide knowledge of current physics that an erudite paper by a 'lady of high rank', and another by one of France's best poets, would sit well in its *Académie Memoires*, along with the three winning papers. This meant that Émilie was about to become the first woman to have a scientific paper published by the prestigious Parisian Academy of Sciences.

*Chapter 8*

# SEARCHING FOR 'ENERGY': ÉMILIE DISCOVERS LEIBNIZ

Émilie's and Voltaire's dissertations on the nature of 'fire' allow us to see where they disagreed with each other – something that is not evident by studying only the *Elements*. But what particularly fascinates me is the insight these essays give into the way physicists struggled to find the right language simply to discuss everyday notions of heat and light.

The first experimental evidence clearly suggesting heat is not a specific material substance was found sixty years after Émilie and Voltaire submitted their essays. Traditionally, heat had been defined in terms of 'caloric', a material substance that supposedly occurred naturally in every object, although some bodies had more of it than others. Consequently, it was believed that caloric could be transferred to other bodies but it could not be 'created' by mechanical or other processes. By weighing metals before and after heating, Voltaire had hoped to measure the weight of caloric, and to thereby give credence to the analogous idea that light particles, too, had weight. In 1798, however, Benjamin Thompson came up with a completely different idea of heat.

He was working for the Bavarian government, supervising the construction of military cannon, and he noticed the heat produced during the process of boring the cannon continued even when the cutting instrument had become blunt; he made further experiments that led him to conclude the heat was produced not by the cutting itself – which had been assumed to release particles of caloric in the process of subdividing the metal – but by simple friction. More specifically, he suggested that the heat in the metal arose from the *motion* of its particles caused by friction – because as long as he kept boring, heat kept on being produced. By contrast, if the metal contained a specific amount of natural 'caloric', then it would eventually run out and the boring process would no longer produce heat. It would take almost a century for Thompson's inkling of the notion of kinetic energy to be adequately formulated, and for the relationship between mechanical energy and heat energy to be quantified, but he had paved the way for the modern interpretation of heat in terms of the increased motion of molecules in the heated matter, rather than the addition of a separate, distinct substance.

Although conclusive experiments on the nature of energy would not take place until the late nineteenth century, theoreticians like Leibniz had been groping towards the idea since the end of the seventeenth century. He had defined the 'force' of a moving object in terms of what we now call kinetic energy, but his work was controversial because his concepts were rudimentary, and so was the language available to describe them.

In particular, Leibniz's ideas sparked confusion over the difference between 'kinetic energy' and 'momentum', although these terms had not yet entered the formal vocabulary of physics. Instead of the term 'momentum', Descartes had spoken of the 'quantity of motion' of a moving object, which he described intuitively as its 'bulk' multiplied by its velocity. Newton made this idea more precise in the *Principia*, when he replaced the vague term 'bulk' by 'quantity of matter' or 'mass', which he defined in terms of relative density or, alternatively, of relative weight. In fact, as I implied earlier, he was the first to distinguish clearly between the concepts of mass and weight. With the concept of mass defined, the intuitive Cartesian term 'quantity of motion' now had a formula: mass times velocity, which can be written symbolically as $m \times v$ (or, more simply, $mv$). But Leibniz had an additional formula: mass times the *square* of the velocity, or $mv^2$, which evidently defined a different quantity, which he called *vis viva* or 'living force'. (The modern formula for kinetic energy is ½ $mv^2$, as I mentioned earlier, but it is only a difference of scaling.)

Leibniz chose the name 'living force' because he believed there was some animating, internal power associated with an object in motion, but in his day – and in Émilie's, forty years later – there was confusion as to whether or not this quantity existed at all. After all, Leibniz was trying to articulate an entirely new concept. But his choice of language was provocative and ultimately counterproductive. He assumed 'living force' was acquired after a body had been moving for a while so that it was in 'full flight' and therefore full of 'life force'; by contrast, he used the term 'dead' force in relation to the Cartesian concept of momentum, which he described as the motion a body acquired at the very beginning of its journey, immediately after it received the impulsive force that started it moving in the first place. His work was also confusing because in different places he used different terms for the same thing: 'force', 'living force', 'kinetic action', 'active force' and even 'force of energy'. Nevertheless, at the heart of his argument was Galileo's rigorous experimental discovery that the distance an object falls under the force of gravity is proportional to the *square* of the velocity it acquires, not simply to the velocity, so there was no doubt in Leibniz's mind that $mv^2$ referred to a physically meaningful quantity. He described his conception of this quantity in a remarkably simple but entirely intuitive argument, which Émilie would paraphrase in a new project she had begun thinking about: a book called *Institutions de physique* (*Fundamentals* [*or Foundations*] *of Physics*).

> Suppose two travellers are walking equally fast [with speed $v$], and that the first travels one league in one hour, and the other walks two leagues in two hours. Everyone acknowledges that the second traveller has covered double the distance of the first, and that the force he has employed to walk two leagues is double that which the first person employed to travel one league. Now, suppose that a third traveller travels these two leagues in one hour, which means he walks with twice the speed of the others; it is evident that the third traveller has employed two times as much force

> as the one who travelled these same two leagues in two hours . . . Now, since the third traveller [whose speed is $2v$] employed two times as much force as the second traveller, and the second employed two times as much as the first [whose speed is $v$], it is clear that the traveller who walked with twice the speed in the same time employed four times more force, and that consequently the forces that these travellers have expended are as the squares of their speeds.

Clearly, the travellers' 'force' referred to an intuitive idea of the effort or energy they expended *during* their motion, which seemed to be proportional to $v^2$; it was a different thing from momentum, which was proportional to $v$, and which, according to Leibniz, was related to the 'impetus' required to start motion in the first place. (He was correct in that force is described in terms of the change in momentum, whereas the increase in a body's kinetic energy is related to the *distance* it has moved under the influence of such a force. However, a body has both kinetic energy and momentum at any point during its motion – so momentum is just as 'alive' as kinetic energy. But this is hindsight: when Émilie wrote her *Fundamentals*, no-one clearly understood the difference between momentum and 'living force', and the Cartesians did not believe the latter existed at all.)

Several months before Émilie began work on her new book, when she was rushing to finish her eighty-four-page dissertation on fire, she had heard of the *vis viva* controversy, although she had not yet read Leibniz. But she *had* read a more recent paper on the subject by Cartesian Academician Jean-Jacques Dortous de Mairan. He rejected Leibniz's idea, saying the only relevant formula for the motion of bodies was the neo-Cartesian $mv$. At the time, Émilie had admired Mairan's argument, and she praised it in a note in her essay; to her chagrin, she discovered Leibniz's innovative analysis even before the competition results were announced. She particularly regretted having made mention of Mairan's view in her essay because when she applied Leibniz's formula to particles of light, she found dramatic support for her hypothesis that fire and light have no weight. Her argument was based on the fact that the speed of light is so incredibly fast that if particles of light had weight, then their impact would destroy the earth.

The speed of light had been measured for the first time in 1675, by Olé Roemer, and a more accurate measurement had recently been made by James Bradley. This enabled Émilie to estimate that, 'in round numbers', the speed of light was 1,666,600 times that of a cannon ball weighing one pound (assuming 'half a pound of powder' was used in the explosion to eject the cannon ball). Then, using the formula $mv^2$ for the 'force' of impact, she calculated that even if the mass of a light ray were 2,777,555,560,000 times smaller than that of the cannon ball, it would have the same impact.

Alternatively, if the corpuscles were even smaller and our eyes were not destroyed by their impacts but progressively absorbed these millions of corpuscles over weeks and months and years, then our eyes should eventually become heavier, which clearly does not happen. Either way, she concluded, 'I believe it is rigorously demonstrated, by the way we see, by the [speed] of light, and by the laws of impact of bodies, that . . . fire [or light] has no weight, or if it does, we are not able to perceive it.' It was certainly an ingenious argument – and the modern view of light is that it does, indeed, have no intrinsic mass. (However, if a light photon is absorbed by a material body, the body does actually gain an infinitesimal amount of apparent mass from the energy of the light, courtesy of Einstein's $E = mc^2$.) Émilie's argument could also be applied, albeit less dramatically, with $mv$, the formula she had initially favoured. Consequently, she had stronger reasons than this to support Leibniz's formula, and she would spell them out in the new book she was writing. It was another secret project, of which more shortly.

Knowing her essay on fire was to be published, Émilie had written to the Academy asking for her note on Mairan's rejection of *vis viva* to be deleted from the published version of her paper, but her request was denied. (The calculations I quoted above are from a later expanded version of her essay.) She told Maupertuis she was 'angry at this severity', but she was also angry with herself for her earlier ignorance of Leibniz's work. She sounded quite depressed in her letter to

Maupertuis, because she could not keep up with what was going on in the world of science: 'Life is so short, so full of duties and useless details that, having a family and a house, I can hardly ever depart from my little plan of study to read new books. I am in despair at my ignorance. If I were a man, I would be at Mont Valérien with you, and I would dump all the useless things of life. I love study with more fervour than I loved society, but I have realised it too late. Keep your friendship for me – it consoles my pride.'

Émilie had servants, of course, but her domestic 'duties' included overseeing the running of the Cirey estate while her husband was away, managing the children's health and education – sometimes teaching them herself – and using her connections to help friends and family in need: *noblesse oblige*. And then there were Voltaire's ongoing dramas. A new one had just surfaced – a personal attack on Voltaire by his old enemy, abbé Desfontaines – which added to Émilie's depression. Voltaire had come down with a fever – he was always sick when his literary reputation was under attack – but this time his critic had not only attacked his work, but had also accused him of impiety. Desfontaines also accused Voltaire of hubris, which was perhaps a more apt charge, and had the issue of impiety been left out of it, he would have been able to shrug off the attack, reducing it to a witty aphorism.

At a time when people were still occasionally tortured and executed for heresy, Voltaire had reason to worry, and he wrote to Prince Frederick, 'My only protectors are you and Émilie . . . I am in France [only] because Madame du Châtelet is here. I do not hate my country . . . but I wish I could study with more tranquillity and less fear.' Combating Desfontaines's charge would take up much of Émilie and Voltaire's time for the next few months.

Meanwhile, in December 1738 – the time of her thirty-second birthday – Émilie was still worrying about her 'stupid' note on Mairan. 'This will cure me of speaking about things I don't know.' She now had no recourse but to ask the Academy to add an 'erratum' to the essay, disclaiming Mairan's view in favour of Leibniz's, because '[I don't want] to see published in my work something that is contrary to what I now believe'. She was right to be concerned, because it was too easy to see such a change of heart as a weakness on her part, and a typically female weakness at that: writing in the *Encyclopaedia Britannica* almost a century later, John Playfair would praise her translation and commentary on the *Principia*, but when speaking of her writings on the *vis viva* dispute, he added, '[F]rom the fluctuation of her opinions, it seems as if she had not yet entirely exchanged the caprice of fashion for the austerity of science.'

However, when the essays on fire were finally published, Émilie's was highly praised throughout the international Republic of Letters. Maupertuis was one of its admirers, as was Voltaire's friend Prince Frederick, who would become king of Prussia in 1740, after which he soon became known as Frederick the Great. He was already in the process of courting the best writers and scientists in Europe, including Voltaire, Maupertuis, Euler and Algarotti, in order to build the prestige of the Academy at Berlin, which had been established in 1710 at the initiative of Leibniz. Frederick wrote to Émilie, 'Without wishing to flatter you, I can assure you that I should never have believed your sex, usually so delightfully gifted with all the graces, [would be] capable also of such deep knowledge, minute research, and solid discovery. Ladies owe you . . . !' He wrote in a similar vein to Voltaire, 'I can only say I was astonished when I read it. One would never imagine that such a treatise could be produced by a woman. Moreover, the style is masculine and in every way suitable to the subject.'

Not everyone was so enthusiastic, of course – especially not Mairan. As a member of the Academy of Sciences, he had seen Émilie's original essay; consequently, when he saw the published version – with the erratum disclaiming her initial praise of the paper in which he rejected the idea of *vis viva* – he was understandably upset. But he let it pass – he was a modest man, not overly ambitious or proud. He did not know that Émilie was already writing a book that would inadvertently bring the *vis viva* debate firmly back onto the international scientific stage.

*

Émilie had begun her *Fundamentals of Physics* with the intention of presenting a deeper discussion of Newton's work than she and Voltaire had given in the *Elements*. After her discovery of Leibniz, however, she decided to broaden her scope, and to combine her detailed description of Newton's physics with an analysis not only of *vis viva* but also of Leibniz's entire philosophy. Like Descartes's, Leibniz's work today often seems little more than a footnote in the history of science, given Newton's triumphantly successful paradigm of theoretical physics. But Leibniz was an extraordinary figure, and Émilie's *Fundamentals of Physics* gives a fascinating perspective on the relationship between his ideas and Newton's, as perceived in Continental Europe in the 1730s and 1740s when Newton's view did not yet prevail.

As with all her scientific work, Émilie's new book had to be fitted around her inevitable domestic upheavals. In May 1739, she and her husband moved their household to Brussels, taking not only the two children, but also Voltaire and the Swiss mathematician Samuel König, whom Émilie had recently hired, on Maupertuis's recommendation, to tutor her in higher mathematics. She and Voltaire would be based in Brussels for the next two years, while Émilie attended to an inheritance case of her husband (who had officially given her the legal right to do this – as a woman, she had no legal authority herself). The reason for such a dramatic move was that the Du Châtelets were far from the wealthiest of the nobility, and the contested Belgian inheritance would do a great deal to improve their circumstances (which sometimes fluctuated to such an extent that their furniture was sold to pay debts, or else, like many others of their class, they simply left debts unpaid). The marquis would return to Brussels occasionally, when his military duties allowed it, but Émilie took charge of the matter, with the help of Voltaire and a lawyer. She took her role so seriously that she spent much effort teaching herself both law and Flemish.

Somehow, she also managed to find time to study with König, to work on her *Fundamentals*, and even to make trips back to Paris – to go to the opera, to visit with friends like Clairaut and Maupertuis, and to spend time with her dying friend, Madame de Richelieu, who would finally succumb to tuberculosis in 1740.

During an earlier sojourn, in October 1739, Voltaire wrote the young philosopher Claude-Adrièn Helvétius that Émilie 'has brought with her to Paris her König, who has no imagination or sense, but who, as you know, is what they call a great metaphysician . . . He swears, following Leibniz, that *extension* is made up of *non*-extended monads, and that *impenetrable* matter is composed of tiny *penetrable* monads. He believes that each monad is a mirror of his universe. When one believes all that, one believes in . . . miracles.'

This was a reference to the philosophical foundation of Leibniz's metaphysics, in which hypothetical 'monads' were the ultimate, irreducible building blocks of the universe. But monads were not material 'atoms', so they were not subject to any of the ordinary, external forces of nature; nor did they have extension, mass, or any other material quality. Rather, they were metaphysical essences, and each created being in the universe had a unique, dynamic monad as its blueprint. Furthermore, according to Leibniz, 'Every individual substance expresses the whole universe in its own manner', including 'all its experiences [and . . . ] the whole sequence of exterior events'. It was quite an elaborate vision, but it was purely qualitative, in contrast to Newton's precise, quantitative theorems and predictions, which Voltaire admired so much. Émilie admired Newton's method, too, but in the process of researching *vis viva*, she had become entranced by the way Leibniz had tried to extend Descartes's vision of a completely unified system of knowledge – secular and spiritual, physical and metaphysical. Such a unified vision had, and still has, a certain appeal, and it certainly struck a chord with Émilie.

The term 'monad' (or equivalent) had already had a long history, beginning with the Pythagoreans, those famous ancient Greek mathematical mystics for whom 'The Monad' or 'The Indivisible One' was the first existing being, which gave rise to the dyad, and then to all numbers, points, lines, shapes and so on, in increasing order of geometrical complexity. Lines and shapes have 'extension', and so they can generate the material objects of the natural world; Voltaire's mention of 'extension' and 'impenetrability', in his letter to Helvétius, referred to these long-recognised fundamental qualities of solid objects. The Pythagorean monad was adapted over the millennia by many philosophers seeking to explain the mystery of existence. In developing his particular idea of monads of 'vital active force', Leibniz had been influenced by the work of several such predecessors, including his English contemporary, the seventeenth-century philosopher, Lady Anne Conway. (Like Émilie's Cirey, Conway's home at Ragley Hall in England had been an important gathering place for intellectual discussion.)

Voltaire's extreme loyalty to Newton meant that as far as he was concerned, you were either for or against Newton, and he interpreted Émilie's interest in Leibnizian metaphysics as a betrayal. Besides, there was the undeniable fact that Leibniz had made a mistake by rejecting Newton's theory of gravitational attraction simply because it did not address the cause of gravity itself. In her new book, however, Émilie was trying something very different from Leibniz and others who rejected Newton's theory of gravity on metaphysical grounds: she was attempting to integrate the Newtonian paradigm within a larger philosophical framework.

In her *Fundamentals*, she praised Descartes for his philosophical method of clear conceptualisation and systematic doubt, but as I mentioned earlier, she believed he had pushed his method too far: she believed reason alone, without experiment or rigorous mathematical proof, was too subjective. In the *Elements*, she and Voltaire had contrasted Descartes's method with Newton's, but now she regarded both Newton and Leibniz as embodiments of the union of mathematics and experiment. And she felt Leibniz had the advantage, in that his philosophical framework was more general, because it extended beyond physics. Consequently, she advocated the philosophical principles that Leibniz had used – notably, the principles of contradiction (something cannot simultaneously be and not be), continuity (all change is continuous), and sufficient reason (which Leibniz defined as: 'a reason can be given for every truth, or, as is commonly said, nothing happens without a cause'). These common-sense principles are fundamental to mathematics and science as well as philosophy, and to some extent they had been used by philosophers long before Leibniz.

As for monads, Émilie admitted the idea of these non-material, indivisible constructs 'stuns the imagination': with no extension, size, shape or weight, they cannot be represented by images, but can only be conceived by the mind. As he indicated in his letter to Helvétius, Voltaire thought it ridiculous to explain *material* properties (like 'extension') in terms of a *non*-material substance, but in preparing her *Fundamentals*, Émilie looked more deeply into the logic behind Leibniz's idea. She was especially interested in the way Leibniz applied his principles to critique Newton's theory of gravity. She also found value in asking metaphysical questions about the limits of what is knowable, and at that stage, she felt Leibniz's principle of sufficient reason was 'a compass that can lead us through the shifting sands of this science'.

Émilie based much of her analysis on the work of Christian Wolff, who had studied under Leibniz himself; Wolff had collected and systematised his master's scattered and often unpublished writings, an endeavour that had made him the world's chief authority on Leibniz's work. Following Wolff, Émilie explained the reason for monads with an analogy: if you are asked, 'Why does a watch work?' and you are told, 'Because it is a watch,' then you are none the wiser; to understand the working of a watch you need to know about springs and wheels – that is, you need to know about objects that are *not* watches. Similarly, if you ask why does matter have extension and you are told, 'Because it is made of atoms, each of which has a tiny extension,' then this is the same as saying, 'There is extension because there is extension.' Material atoms are not, therefore, a sufficient reason for the existence of extension. In other words, it was tautological to explain extension by invoking the then-hypothetical existence of material atoms (which would not be discovered experimentally until the eve of the twentieth century). Consequently, Leibniz had argued there must be a different kind of fundamental building block – one *without* the qualities of extension and impenetrability of matter.

It was, indeed, a logical argument, but it was logic predicated on such a narrow set of premises that Voltaire's instinctive rejection was not without foundation. For a start, the idea of 'extension' as the primary definition of matter was limiting and old-fashioned, given Newton's new concept of mass, but Leibniz's rejection of atoms on the grounds they were tautological assumed there was no other way to describe matter than extension; it also assumed we *should* be able to logically describe such matter, or at least to give a 'sufficient reason' for its existence. Newton, on the other hand, admitted there are some things physicists cannot describe, at least for the moment. His concept of mass did not require any assumption about the internal structure of matter.

Nevertheless, in terms of Leibniz's strict, *a priori* philosophical assumptions, there was an intellectual rigour behind his rejection of the concept of gravitational attraction, which Émilie's account makes clear. She explained that because Newton's theory lacked an explanation of the cause of gravity itself, Leibniz considered the argument that the planets move because of gravitational attraction towards the sun was like saying the planets move because they move. And in a way, he was right. But Émilie and Voltaire both realised that Leibniz – along with the Cartesians – had failed to see that Newton *agreed* gravity's unknown cause was ontologically problematic, but that philosophically, he was more interested in certainties than possibilities, which is why he built his theory not on hypotheses about possible causal mechanisms, but on observable physical effects.

However, Émilie did agree with Leibniz's proof that gravity was not an inherent part of matter. She began her argument by recapitulating Leibniz's rejection of Locke's idea that perhaps matter without a soul can think. Following Wolff, she formulated this as a logical argument along the following lines: everything that exists must have its own sufficient reason for being – that is, its own essence. The 'doctrine of essences' assumed essential attributes are not communicable, so the attributes of one type of 'thing' cannot be transferred *arbitrarily* to another. For example, the attributes of solid matter – extension and impenetrability – are founded in the essence of matter, and are quite different from the immaterial essence of thought. This means the immateriality of thought cannot be transferred to matter, and so matter (such as a purely material brain) cannot think. Of course, Locke would have pointed out correctly that this is a circular argument, because it assumes without proof that the essential nature or mechanism of thought is immaterial.

Descartes, too, had underestimated the difficulty of proving assumptions in logical arguments, despite his attempt at rigour through 'clear conceptualisation'. For example, in trying to prove the soul is immortal, he noted that living bodies are composed of many separate parts, the implication being that such bodies can therefore decompose after death. By contrast, he claimed, the soul is a whole entity, not a composite one, 'because we cannot conceive of half a soul'. Therefore, he concluded, the soul is not subject to decomposition and so it is immortal. The conclusion may well be true, but Descartes had not proved it, any more than Leibniz had proved the immateriality of thought.

Circular logic notwithstanding, Émilie utilised a 'doctrine of essences' argument to 'prove' that gravity could not be inherent in matter. She then went on to 'prove' that gravity is not a sufficient reason to be the *cause* of planetary motion – rather, it is simply a useful *explanation* of the observed facts. (This last comment has some validity because, to some extent, the issue here, and in much of theoretical physics, is one of language, as I mentioned in connection with centrifugal and centripetal force.) Émilie was captivated by the way conclusions follow logically from Leibniz's assumptions, which are so few in number and so tight in conception that arguments proceed from them in an economical and satisfying way. Logic is still an important part of philosophy today, but as I indicated earlier, it stresses the need for correct premises, whereas some of Émilie's Leibnizian arguments are unconvincing because of their unproven assumptions.

She was on firmer ground when she discussed the nature of time and space. Like Leibniz, she believed space and time were merely conventions for the *relative order* of one body with respect to another and one event in relation to another another. By contrast, Newton's attempt to define motion in the simplest possible way so that it could be analysed mathematically meant he had been forced to assume, at least in mathematical terms, that time and space were absolute, the same for everyone no matter where they are in the universe or how they are moving. He realised this was an assumption – something he had tried assiduously to avoid – and so his experimental measurements were all relative. But his mathematical analysis required a fixed coordinate system for space and time, and it is an ironic validation of the Newtonian method that, for all his attempts to avoid using unproven assumptions, in a sense it was his subtle hypothesis of absolute space and time that ultimately led to his theory of gravity being superseded by Einstein's relativistic one.

Einstein said that in the contemporary state of physics at that time, Newton's reluctant use of absolute space and time coordinates was 'the only possible [choice], and particularly the only fruitful one'. But, he added, 'The resistance [to Newton's choice] by Leibniz and Huygens, intuitively well-founded but supported by inadequate arguments, was actually justified . . .' Over the next two centuries, many philosophers (including Hume, Kant and Mach) had continued the debate, but it was Einstein who developed the modern mathematical theory of relativistic space and time. (Poincaré had come close, but he had used the circular Cartesian method of developing a theory from assumed causes, whereas Einstein derived his theory from physical principles in the Newtonian style.) Back in 1739, Émilie gave a perceptive account of relative motion in which she also acknowledged the usefulness of absolute space and time in 'everyday discourse', as she put it, both verbal and mathematical.

As for metaphysics in general, modern physicists tend not to worry about the metaphysical distinctions that bothered Leibniz and his colleagues; rather, they follow Newton in accepting that physics can only deal with the most basic of physical processes. Nevertheless, metaphysics still has a place in the creative *process* of theoretical physics. After all, Leibniz's metaphysical speculations on *vis viva* laid the foundations for the law of conservation of energy (of which more later), while some physicists wonder if, at the quantum level, ideas like monads and Cartesian vortices might provide helpful preliminary models for visualising some of the bizarre behaviours of elementary particles. And philosophers still occasionally argue about the principle of sufficient reason, although most believe quantum mechanics violates it. For example, in the famous 'two-slit' experiment, particles like electrons or photons are fired at a screen with two tiny slits in it. In some sense, these particles seem to 'know' whether one or both slits are open, because collectively they behave entirely differently in each case. For a rough analogy in everyday terms, imagine posting a batch of letters into a letter box containing either one slit or two side-by-side slits: in the first case, you would get a single pile of letters at the bottom of the box, and in the second case, you would get two overlapping piles. In the quantum case, however, you would get the equivalent of a single pile of letters when there is only one slit, and a neat arrangement of four or five distinct piles when there are two slits side-by-side. I will discuss the implications of this conundrum in Mary Somerville's story, but the point here is that this bizarre behaviour defies explanation in terms of a mechanical, Leibniz-style 'sufficient reason' (or, indeed, in terms of Newtonian physics).

Nevertheless, some philosophers still look to traditional metaphysics, including variants of the principle of sufficient reason, in the hope that 'all these intractable quantum paradoxes might just dissolve into sheer Necessity', as Allan Randall put it recently – which is interesting language in light of Émilie's view not of metaphysics but of the theory of gravity. She wrote in her *Fundamentals*, 'All the astronomical phenomena that appeared virtually inexplicable in the system of vortices, seem to be simply necessary corollaries of [the theory of] universal attraction . . .' If she had been prepared to leave it there, without giving the last word to Leibniz's rigid form of the principle of sufficient reason, she would have anticipated the position of many philosophers and physicists today, for whom mathematics itself has become the new metaphysics. For instance, Leibniz's various monads supposedly contained the blueprint of everything that happens in the universe, but it could be said that Newton's law of gravity contains the blueprint for planetary motion, while Einstein's equations of relativity contain the 'history', past and future, of the universe itself.

Émilie did not articulate such a modern position – partly because she wanted a philosophy that took account of the history of humanity as well as that of impassive celestial spheres – but she went on to say the idea of gravitational attraction explains material phenomena 'marvellously well': unlike Leibniz, she did not allow the principle of sufficient reason to prevent her accepting Newton's theory. She simply stipulated that 'attraction is not an inherent property of matter', and that the explanatory success of attraction should not divert philosophers from also seeking the cause of gravity itself. Newton would not have disagreed. She then proceeded to give a more detailed account of Newton's theory of gravity than in the *Elements*, including the extraordinary chain of reasoning that led Newton to the inverse-square law and the idea of universal gravity.

She also discussed Leibniz's key contribution to physics (rather than metaphysics), namely his idea of *vis viva.* Recall that the dichotomy at the heart of the dispute opposed 'living force' ($mv^2$) to 'quantity of motion' or momentum ($mv$). The Cartesians believed there was only one type of force, the push-pull direct impact force, which they assumed was the only way to give motion to an object. They also believed that the only meaningful measure of the amount of motion imparted by such a force was momentum, mass times velocity. They did not consider the possibility that an object could acquire 'living force' as well as momentum, because they did not believe 'living force' was a real quantity.

In the very last chapter of her *Fundamentals*, Émilie discussed various arguments by Leibniz, Hermann and others (including herself), which were designed to 'prove' that 'living force' existed, and to illustrate the difference between it and momentum. She also confidently discussed flaws in the arguments of Leibniz's critics, like Mairan, Jurin and Maclaurin. From a modern perspective, however, the debate at that stage was confusing and inconclusive, because the language did not yet exist to adequately formulate Leibniz's ideas. The physical relationships between force, momentum and kinetic energy would not be fully understood until the nineteenth century, so I will further explore Leibniz's legacy – particularly with regard to the principle of 'conservation of energy' – in Mary Somerville's story. The important point here is that Émilie's analysis of such a controversial subject showed that she saw herself as a scholar to be taken seriously.

## *Chapter 9*

# MATHEMATICS AND FREE WILL

For all that she was entranced by philosophy, Émilie found the logic of mathematics even more seductive. She told Maupertuis she 'loved' mathematics for its rigour, and in her *Fundamentals of Physics* she spoke not only of the 'marvellous' explanations afforded by the mathematics of gravitational attraction, but also, for example, of the 'beautiful' property of a cycloid, which makes this curve the unique path by which an object will descend from one point to another in the quickest possible time.

This 'path of quickest descent' does not refer to an object that is simply dropped from a height and falls vertically downwards, but rather to a constrained 'particle' like a bead or ball bearing sliding under the influence of gravity down a smooth curved surface like a chute or slide. That the profile of this 'optimal' surface should be part of a cycloid is 'surprising', Émilie noted, because you would think the quickest route would also be the shortest – that is, a straight line between the given points at each end of the slide. (If the speed of a sliding object were constant, then the shortest path would indeed be the quickest. But objects pick up speed as they fall, so the optimal path is the one that allows the speed of a sliding object to change in just such a way as to minimise the total time taken.) Émilie explained that the solution to the problem is actually *part* of an *inverted* cycloid, where the cycloid itself can be visualised as the curve generated by a point marked on the rim of a 'wheel of a coach' as the wheel rolls along a straight road. In other words, a cycloid is a series of arcs that look roughly semicircular but which have their own precise mathematical form.

After noting that Galileo had first proposed this problem, and that he had wrongly believed the path of quickest descent was part of a circle, Émilie said Bernouilli was the first to solve it correctly, and that subsequently all the great men of science had solved it in various ways. She mentioned that Leibniz's solution referred to the principle of sufficient reason, but she did not mention Newton's elegant solution. Her emphasis on Bernouilli and Leibniz suggests she had heard only Bernouilli's side of the story – a story that centres on the notorious calculus priority dispute between Newton and Leibniz.

The gist of this dispute is well known, but it is worth repeating that it was not something either of the protagonists had wanted – their followers had initiated it. Bernouilli was a leading partisan on Leibniz's side, but Newton had his own 'toadies', as Bernouilli called one of Newton's more aggressive supporters (John Keill). History's verdict is that Newton invented calculus first, but that Leibniz's discovery was essentially independent, and he has priority of publication – Newton was notoriously secretive about his work, but astonishingly his early papers on calculus were actually rejected by both the Royal Society and Cambridge University Press. However, both men did become personally embroiled in the dispute, because of accusations of plagiarism made against each of them: Newton famously used his position as president of the Royal Society to influence its decision on the issue, but Leibniz, too, behaved less than impeccably, as the following story illustrates.

It began back in June 1696, when Bernouilli published, in the scholarly journal *Acta Eruditorum*, a challenge to the international mathematics community to find this 'path of quickest descent'. Bernouilli called it the 'brachistochrone problem' (pronounced 'brack-ist-o-krone') – from the Greek words 'brachistos', meaning shortest, and 'chronos', meaning time. He had already discovered that its solution is part of a cycloid, but he hoped to show his mathematical superiority by giving a limit of six months for the submission of solutions. By the end of the year, he must have been gratified by the fact that he had received no satisfactory replies, but he did receive a letter from Leibniz, saying he had solved the problem and asking Bernouilli to republish the challenge and to extend the due date until Easter. Bernouilli did so, and he also sent the problem directly to Newton.

There is suggestive historical evidence that Bernouilli, and presumably Leibniz too, assumed Newton had not been able to solve the problem, because he had not submitted a solution when the problem was first proposed. But Newton had not answered the initial challenge – he may not even have been aware of it – because he was then taking a break from mathematics. He used to work sixteen-hour days seven days a week for months on end and, not surprisingly, he had had a nervous breakdown in the early 1690s. (Incidentally, some historians wonder if mercury poisoning from his alchemical experiments also contributed to his breakdown.)

In 1696, Newton took on a position as warden of the Royal Mint, presumably in order to help his recovery by taking up easier work than science; at any rate, he was not thinking much about mathematics. But the calculus priority dispute was gearing up, and when he received his personal copy of Bernouilli's announcement, Newton immediately realised that Bernouilli and Leibniz must have assumed that solving the brachistochrone problem required advanced calculus, and that they were hoping to prove Leibniz's priority in its invention by publicly demonstrating that Newton was unable to find a solution. This would prove he did not have as good a grasp of calculus as Leibniz – or even Leibniz's disciple, Bernouilli – thereby revealing Leibniz as the 'true' founder.

Calculus is the art of manipulating infinitesimal increments of time, distance or any other quantity. Recall that in proving that centripetal motion proceeds according to the 'law of equal areas', Newton used tiny triangles to approximate the sectors swept out by a planet in an 'instant' of time. But the arc of a planetary orbit is a curved line, not a straight line as in a triangle, so the use of triangles to approximate sectors means the planet's orbit is being approximated by a series of tiny straight lines. This simulates motion itself, which proceeds with one tiny step after another – except that unless each step is *infinitesimally* small, the motion will be jerky rather than continuous and smooth. Newton's proof, then, focused on the area of a sector swept out in an infinitesimal time interval, and he then 'added up' these tiny areas to find the larger areas swept out during the planet's smooth, continuous motion. This concept of 'adding up' infinitesimals is the basis of 'integral' calculus. (By contrast, 'differential' calculus is the art of manipulating ratios – or, more correctly, 'rates of change' – involving infinitesimally small quantities. For example, speed is the ratio of distance and time – in units such as metres per second or kilometres or miles per hour – so the *instantaneous* speed of a moving object can be thought of as the ratio of the distance travelled in an infinitesimal interval of time, as measured at the required 'instant'.)

The fact that Newton used geometry to find the areas of his planetary sectors illustrates why there was doubt about his facility with calculus – because the true power of calculus lies in its *algebraic* algorithms, which Newton rarely used overtly in his masterwork. However, geometrical constructions underlie the algorithms of calculus, as modern high-school students of calculus know (and as you can see in the Appendix), although today the derivation of these algorithms relies upon the sophisticated concept of a 'limit' that was not made rigorous until the nineteenth century. A key reason it took nearly two hundred years to do this is that at the heart of calculus there is a very tricky question: how do you quantify an 'infinitesimal'? Such a quantity is 'not quite' zero, but how do you decide when you are 'close enough' to zero? In the Appendix, I have given a simple example of why this was such a confusing question. Nevertheless, intuition generally proved to be a good guide to the handling of infinitesimals, which is why Newton and Leibniz had been able to come up with their algorithms in the first place. But the lack of a precise, *mathematical* 'limit' concept meant that Newton preferred to use more transparent geometrical arguments if possible.

Leibniz, too, was aware of the problems, although he did not appear to understand their significance as well as Newton did. In the absence of mathematical rigour, he seems to have relied on a metaphysical justification, likening his infinitesimal quantities to monads. But Leibniz also created an extremely user-friendly notation for calculus – an algebraic symbolism so well adapted to the intuitive concepts of calculus that it could be used relatively effortlessly. (In keeping with this 'automated' approach, he also drew up a plan for the first binary computer, but it was never built.) Given the historical importance of Leibniz's version of calculus compared with Newton's – both Émilie and Mary would use Leibniz's version, which we still use today – I have given in the Appendix a brief indication of the benefit of Leibniz's notation over Newton's.

In fact, by the early eighteenth century, Leibniz's symbolism had already become widespread on the Continent, through the work of Leibnizian disciples like Bernouilli, and later through Bernouilli's protégés, Maupertuis, Clairaut and Euler, and also *their* students, notably Émilie – all of whom began the process of translating various geometric proofs in the *Principia* into Leibnizian language. (Émilie would do this specifically in her later commentary on the *Principia*.)

The notational utility of Leibnizian calculus was another reason that led Bernouilli and many others to suppose Leibniz understood calculus better than Newton did. Bernouilli revealed this opinion in the wording of his second public brachistochrone challenge, in which he said that, when published at Easter, his and Leibniz's solutions would reveal 'the narrow limits of the common geometry'. Not that Bernouilli named Newton; rather, he confidently announced there would be few who could solve 'our excellent problem', few even among 'the very mathematicians who boast that by the remarkable methods they so greatly commend, they have not only penetrated deeply the secret places of esoteric geometry, but have also wonderfully extended its bounds by means of golden theorems which (so they thought) were known to no-one, but which, in fact, had long previously been published by others.' (Recall Leibniz had published his calculus before Newton did.)

The calculus priority dispute had nationalistic overtones, and when he read Bernouilli's challenge, Newton was furious at being 'teased by foreigners', as he put it. Although Bernouilli and Leibniz had allowed three months from the time he received his copy of the problem, he took great delight in solving it overnight. He submitted his solution anonymously to the president of the Royal Society. He did not show the underlying details of his method, and it has been argued he probably used geometry rather than calculus (although he certainly could have used calculus if he had wanted to). His solution eventually reached Bernouilli, who – much to his embarrassment – immediately recognised Newton's stamp: namely, the uncanny ability to solve complex problems in an intuitive leap. In this case, Newton simply gave a one-paragraph statement plus a diagram, in which he took an arbitrary cycloid and showed how to derive from it the correct, minimising cycloid connecting the two given points.

Leibniz, too, had solved the problem overnight, but now he was mortified by his complicity in Bernouilli's attempt to trick Newton, and he wrote to the Royal Society distancing himself from the genesis of the challenge. Apparently, Émilie did not know any of this history, but in her *Fundamentals*, she was very taken with Leibniz's way of thinking about the brachistochrone problem, because it involved an all-embracing philosophy that seemed to unite such disparate subjects as mathematics, physics, the existence of evil and even the nature of free will.

Free will was a tantalising topic for free thinkers in a repressive society, and it had fascinated Émilie since she was a teenager. Newton's precise mathematical laws of nature seemed to suggest the universe ran according to a predetermined design, but Leibniz's monad philosophy, too, implied everything was predetermined by God. This was because Leibniz's monads, his fundamental life-giving essences that were the building blocks of his universe, contained the seeds of all the predetermined events of the entire universe, from its conception to eternity.

To solve the problem of freedom in such a predetermined universe, Leibniz argued God had had a free choice in manifesting the world that actually does exist – and that he chose the best of all the *possible* worlds. This freedom of choice was supposedly reflected in human reason, because people always seek the best possible outcomes for themselves. It was a somewhat contrived argument for the existence of human free will, but Leibniz also used it to explain the existence of evil in a world supposedly created by a benign God. Émilie supported Leibniz's idea, in which the best possible world did not mean that this was a perfect world – rather, it meant that 'an imperfection in the part may be required for a greater perfection in the whole', in Leibniz's words, and that evil may accompany a 'greater good'. He gave a simple illustration: 'An army general will prefer a great victory with a slight wound to a condition without wound and without victory.' Émilie also interpreted Leibniz's argument to mean that the physical laws of the universe were those that enabled the universe to run efficiently, in 'the best possible way'.

Voltaire and Émilie disagreed on this matter, and later he would famously satirise Leibniz's position, in his novel *Candide*. Today, some commentators accuse him of confusing the phrase 'best possible world' with 'perfect world', and of simplistically poking fun at Leibniz's philosophy; on the other hand, his treatment of the Lisbon earthquake of 1755, which features in both *Candide* and his moving *Poem on the Lisbon Disaster*, suggests he was using his powerful literary skill not merely to scoff but to strenuously reject the use of philosophy (and religious dogma) to explain away suffering. (*Candide* also satirises the real-life burning of 'heretics' that followed in the wake of the earthquake in the vain hope of gaining protection from God against a second 'after-shock' quake.) But Émilie was not the only *philosophe* who supported Leibniz's attempt to find a philosophical framework to 'explain' why a loving God would allow evil in the world: for instance, Jean-Jacques Rousseau argued that Voltaire had simply replaced Leibniz's best-possible-world 'optimism' with excessive pessimism.

Émilie was also fascinated by the *mathematical* ramifications of Leibniz's philosophy, and in her *Fundamentals* she explained how his solution of the brachistochrone problem fitted into his philosophical framework. The unique 'path of quickest descent' was part of God's plan to give a sufficient reason for everything that actually manifests in this best of all possible worlds, as opposed to all the other possibilities that do not. In this case, the 'sufficient reason' was economy of time: out of all the possible paths that a sliding object could take, the cycloid is the one that requires the least time. Other paths might have other sufficient reasons: notably, a straight line gives the shortest distance.

These two particular 'sufficient reasons' have obvious applications in engineering designs, but Émilie showed they also govern natural processes, such as the refraction of light. Her explanation is connected with the ancient metaphysical principle that nature acts as simply as possible. Aristotle had articulated this idea over two millennia ago, and four centuries later (in the second century of the common era), Heron of Alexandria used it to explain the law of light reflection. Recall this simple law of nature is that the angle of incidence of a light beam as it hits a reflective surface has the same magnitude as the angle at which it is reflected from the surface. Heron used simple geometry to show that such a reflected light ray – as it leaves its source and then is reflected off a mirror and into an observer's eye – takes the shortest possible path.

Fifteen hundred years later (and almost half a century before the *Principia*), Pierre de Fermat applied a similar analysis to the refraction of light. As Émilie noted, Fermat showed that the actual path taken by a light ray that is refracted at a surface is the path that light can travel in the shortest possible *time*. Minimising time rather than distance was an important insight, she explained, because clearly, the path a light ray travels between a point in the air and a point inside the refracting medium is not the shortest in *length*, otherwise it would be a straight line rather than a bent one. (Just as in the brachistochrone problem, if the speed of light were constant throughout the process of refraction, the path taking the shortest time would also be the shortest in length; but light travels at different speeds in different media.)

Bernouilli and Leibniz had adapted Fermat's approach in solving the brachistochrone problem, but Émilie referred the reader to more advanced texts for details, including what she diplomatically referred to as 'M. Mairan's excellent memoir' on Fermat's argument. In the mid 1740s, several years after the *Fundamentals* was published, Maupertuis and Euler would develop Fermat's method into the modern 'principle of least action'.

Today, choosing from many possible paths by minimising (or economising on) some aspect of an object's motion – such as time or distance or energy – has become a mainstay of modern mathematical physics. This is not for *a priori* metaphysical reasons, although it is remarkable that methods like 'least action' have proved extremely effective in finding equations that describe natural processes. It is tempting to say that nature does, indeed, act economically! I will return briefly to this topic in Mary Somerville's story.

It was König who instructed Émilie in the subtleties of Leibniz's philosophy, but he never settled in to her household. The relationship seems to have been uneasy on both sides: König treated Émilie rather patronisingly, not tailoring his lessons to her needs but proceeding at a fast pace, presenting what he wanted to teach and then asking her to sign the lesson if she 'understood' it. On the other hand, Émilie seems to have regarded him not as a friend and an equal, as with Maupertuis and Clairaut, but as an employee. Like most aristocratic women at the time, she could be rather imperious with the servants: she rightly complained about gender discrimination – her male peers did not have to deal with the prejudices and the daily responsibilities that sometimes overwhelmed her – but she seemed relatively unaware of the problems of class; after all, the French Revolution was still half a century away.

Having long recognised the tension between them, she had already written to ask Maupertuis to persuade Bernouilli's son Jean Bernouilli II to take up the post of live-in tutor – he and Maupertuis had visited Cirey several months earlier, in January 1739, and had given her a taste for calculus. She told Maupertuis, 'I despair of keeping König, and I will feel much easier if you speak to Bernouilli for me.' (Incidentally, the Bernouillis were the greatest mathematical dynasty in history: Jean II also had two mathematical brothers, as well as his famous father, several uncles and a cousin; there had also been his grandfather, and there would be a son and several nephews.)

To help Émilie's case, Voltaire wrote directly to Bernouilli II, assuring him 'of the sincere joy I would have in seeing you with madame la marquise du Châtelet . . . You know the life at Cirey. That of Brussels is more or less the same: it is about liberty, equality and study, with a taste for solitude rather than company.' Unfortunately, the fall-out from the König affair was unreasonably damaging to Émilie, partly because König wildly accused her of plagiarising his lessons in her *Fundamentals*, and Maupertuis felt compelled to dissuade his famous mentor's son from taking up the position. Émilie was heartbroken by Maupertuis's betrayal – and Voltaire went so far as to urge him to apologise: 'A man is always right when he admits he is wrong to a woman. You will regain her friendship, since you still have her esteem.' The rift endured for several months, but eventually Émilie forgave Maupertuis, reassuring him, 'I do not know how to love nor how to be reconciled by halves.' As for her mathematics, she had to persist by herself, asking questions by letter to Maupertuis, Bernouilli and Clairaut.

At the same time as she had been trying to find a teacher for herself, Émilie was also having difficulty finding a tutor for her twelve-year-old son, Florent-Louis. In February 1740, Voltaire wrote to Dutch Newtonian 's-Gravesande to see if he could recommend someone; Voltaire added that it would not matter what university such a tutor had been to, or what his religion was, because 'in the kingdom of Madame du Châtelet, there is absolute freedom of conscience'. Émilie had been trying to find a satisfactory tutor for young Florent-Louis for two years, ever since she had been forced to fire his first tutor, and during that time she had taken over the teaching role herself. In fact, the education of her son was her ostensible reason for writing her *Fundamentals*, which she dedicated to Florent-Louis, although most of the content of the book seems too sophisticated for a twelve-year-old boy.

In the preface to her son, Émilie advised him (and her readers) against partisanship in science: 'It is inappropriate to make a national affair out of the opinions of Newton and Descartes: when it is a question of a physics book, it is only necessary to ask whether or not it is a good book, not whether its author is English, German or French.' She also indirectly acknowledged König's help, telling her son: 'I have drawn from the works of the celebrated Wolff, of whom you have heard much spoken with one of his disciples, who has been at my house for some time, and who sometimes made extracts for me.' (König had studied under Wolff himself.) The indirect nature of this acknowledgment follows from the fact that she published the book anonymously. This was presumably for the same reason she did not include her thirteen-year-old daughter in her dedication: women and girls were not supposed to know much about physics and mathematics.

Nevertheless, it seems surprising that Émilie, who wrote so passionately on the need for women's education, did not choose to educate her daughter solely at home, with her son, but primarily in a convent. Not that young Gabrielle-Pauline was full-time at her convent; in December 1738, for example, Émilie had sent for her to come and play one of the parts in a Cirey theatrical performance, and Madame de Graffigny described with awe how the then twelve-year-old learned her part on the short journey from the convent to Cirey. Graffigny also wrote of Gabrielle-Pauline, 'She is not pretty, but she speaks like her mother, with all the wit and intelligence possible. She is learning Latin, and likes to read – she will not fail her blood.' This reference to Gabrielle-Pauline's aristocratic heritage perhaps explains why Émilie seems, to modern eyes, to have taken less interest in her daughter's education than in her son's: for all her own intellectual efforts to become a thinking person and to rise above her destiny as a woman, she did so from a socially secure position as the marquise du Châtelet. Consequently, she felt her primary duty to her daughter was to arrange an advantageous marriage for her, and there were perceived social advantages for a girl with a pious convent education, especially a superior one such as

Gabrielle-Pauline seemed to receive, since she was learning Latin and literature.

'What pain and care we give every day in the uncertain hope of procuring honours and of augmenting the fortune of our children,' Émilie wrote in the preface to her *Fundamentals.* And, three years later, with the aid of an appropriate dowry, sixteen-year-old Gabrielle-Pauline would, indeed, make a suitable marriage, to the Italian duke, Alfonso de Montenero-Carafa. But neither she nor her brother would carry on their mother's legacy of scholarship. Florent-Louis would, however, attain Émilie's dream of visiting Newton's homeland: in 1767, he would take a post as Ambassador to London, before returning to France as governor of Toulouse and a colonel of the French Guards.

Meanwhile, Émilie was thrilled with the response her *Fundamentals of Physics* received. Wolff was astonished at both her gender and at the 'clarity with which she can talk about the subtlest things', and he and Émilie began corresponding with each other. Her book would be so well received it would be translated into German, the language of Wolff and of Leibniz himself.

In France, the *Fundamentals* was reviewed favourably in scholarly journals like the *Journal des savants*, while the French Wolffian scholar Deschamps responded to the book with 'transports of delight', a sentiment that Émilie and Voltaire's friend Cideville also expressed. Cideville's tribute also highlights the unashamed French penchant for entertaining guests from the boudoir (a habit Mary Somerville would discover with a mixture of shock and delight): 'You are capable, Madame, of awakening a taste for the most abstract sciences . . . Can it be that the sublime author of this grave and dogmatic book is the adorable woman I saw lying in bed three months ago, whose large, fine, gentle eyes, dark eyebrows, charming and noble countenance, ingenious and piquant intellect, cheerfulness and sallies of wit give us all, in truth, quite different things to think of than philosophy?'

Many others in the Academy and in the general Republic of Letters were similarly enchanted by the *Fundamentals*, including those who were not Leibnizians. Clairaut said no-one had made him understand Leibniz's metaphysics as well as Émilie did, while Voltaire told Helvétius he had 'seen nothing on German philosophy that approaches near Madame du Châtelet's book. It is something that is very honourable for her sex and for France' (because there had as yet been no accessible overview in French of Leibniz's and Wolff's works). But the most gratifying response to Émilie's book was surely the long review by Maupertuis, in the *Mercure de France*: 'There appeared at the beginning of this year a work that would give honour to our century if it were by one of the principal members of the Academies of Europe. This work, however, is by a woman, and what makes it even more marvellous, she is a woman who, having been raised with the undisciplined distractions associated with her high birth, has had for a teacher only her genius and her application to self-instruction.'

His thirty-six-page critique attested to the seriousness with which he read Émilie's book despite his lack of enthusiasm for Leibniz's metaphysics: he held the modern view that there was insufficient physical evidence to take Leibniz's 'principles' as axiomatic truths the way Émilie did, although he recognised her skill in explaining Leibniz's ideas. As for the chapters on gravitational attraction, he noted, 'Here the author raises herself well above what she modestly calls *Fundamentals*.'

The book had its critics, of course. Frederick – who was now king of Prussia, and who knew Wolff personally – responded to her with gallantries about her genius and her gender. But he added politely that he felt there were some chapters where the reasoning could have been tightened, particularly on the topic of extension. Privately, he told a friend, 'Minerva now does physics. There is some good in it, since it was König who dictated her theme, which she has adjusted and decorated here and there with words that must have escaped from Voltaire at one of his suppers. But the chapter on extension is pitiable; the order of the work is bad and there are even some mistakes . . . Her friends should have charitably advised her to instruct her son without instructing the world.'

The implicit accusation of plagiarism was unfair: apart from the fact that the *Fundamentals* was about Newton as well as about König's Leibnizian 'theme', Émilie's letters to Maupertuis show that she had first become interested in 'living force', and hence in Leibniz, through Bernouilli and Maupertuis, before she had even met König. Although she also made use of König's 'extracts', she had told Frederick himself that her goal was simply to present Wolff's take on Leibniz to a French audience ('with a French sauce'). She did it so well that many critics, then and now, agree she was clearer even than Wolff himself. Which is not to say the book was perfect in clarity and style – few books are, especially one whose aim was to integrate such a wide range of ideas. But Frederick should have heeded Leibniz and avoided half-baked, egotistical 'carping criticisms': in discussing the longstanding tendency of writers to criticise each other 'too sharply', Leibniz had written, 'Would that people preferred to apply themselves to the task [of making knowledge], rather than waste their time with carping criticisms. They would sacrifice only their vanity!'

Indeed, Frederick was so far from being an expert in physics that Émilie had once offered to give him lessons. His uncharitable response to the *Fundamentals* – in contrast to his earlier admiration of her essay on fire – was a sign of his growing jealousy of Émilie's hold over Voltaire, whom he was trying to entice to move to Berlin. The jealousy was mutual: Voltaire had kept stalling on Frederick's invitation to visit him in Berlin because he did not want to leave Émilie, but when he did finally agree to go, Émilie would later write with relief that Frederick had been astonished that Voltaire returned home to her after only a few weeks. 'I believe [the king] is outraged against me,' she told Argental, 'but I defy him to hate me more than I have hated him these last weeks!'

A more important criticism of Émilie's *Fundamentals* came from Mairan, who had now replaced Fontenelle as the permanent Secretary of the Parisian Academy of Sciences. It was not a position he had sought, but he took seriously his role as the chief defender of French science – so seriously, in fact, that he believed he had to publicly challenge Émilie's support for Leibniz over Descartes on the matter of the correct formula for 'force' ($mv^2$ versus $mv$).

In Mairan's letter in response to the *Fundamentals*, he began by wondering ironically what had happened, since Émilie first submitted her essay on fire, to make her change her opinion on the matter. Then he accused her of not having read his memoir properly, and he patronised her as a woman who did not understand the mathematical subtleties of the argument. In fact, Mairan himself was not a mathematician, and his memoir contained little real mathematical analysis. Clearly he did not expect Émilie to respond – he had experience, education and the Academy of Sciences on his side, and she was merely a self-taught marquise born for a life of frivolous dissipation. But she was not upset by Mairan's attack: on the contrary, she told Argental, 'I don't know if I will reply to him, but I am honoured to have such an adversary.'

In the end, she did respond to Mairan's open letter, and she did so with spirit. Her response included nearly fifty pages of (rather tedious) examples and counter-arguments to refute Mairan's (equally tedious) letter and memoir. This public war of words between Émilie and the most important man in the Academy of Sciences proved she had, indeed, 'arrived' as a scholar to be taken seriously, despite the fact that, as a woman, she could not actually become a member of the Academy. As for the verdict on the winner of the debate, predictably there were mixed reactions, but even reviewers who supported Mairan commented on Émilie's thorough reply.

Although the primary opponents of *vis viva* were the Cartesians, who believed momentum, not *vis viva*, was the true measure of the effect of 'force', many Newtonians – including Voltaire – assumed the argument in favour of *vis viva* was an attack on Newton's ideas. I have mentioned Newton's definition of centripetal force, but in his preliminary 'laws of motion' at the beginning of the *Principia*, he also gave a more general definition in which he defined force as something that causes, and is proportional in magnitude to, a change of momentum. By contrast, Leibniz considered the 'true' measure of force should relate to the change in *vis viva* ('living force'), not momentum. Voltaire, ever on the defensive for Newton, had publicly argued in favour of Mairan and against Émilie. But he told Mairan privately, 'It is a beautiful example for the people of letters, showing that one can be tenderly and respectfully attached to those one disagrees with.' Émilie agreed: she wrote Argental, 'Voltaire's memoir against me must reassure my adversaries.' However, she added that as far as she and Voltaire were concerned, 'One cannot imagine a greater contrast in philosophical sentiments, nor a greater conformity in all else.' Indeed, Voltaire told his friend Cideville that Émilie was the only philosopher he did not mistrust, because 'she is constant in her principles, and even more faithful to her friends than to Leibniz'.

One of the most important aspects of Émilie's debate with Mairan was the fact that, as an outsider, she had had the courage to voice publicly the view held privately by some of the younger, more mathematical members of the Academy of Sciences, like Clairaut and Maupertuis. She was hurt that they did not support her publicly in the dispute, but three years after she had reopened the debate in her *Fundamentals*, Jean Le Rond d'Alembert would explicitly spell out what many of his mathematical contemporaries already knew: both $mv^2$ and $mv$ could be derived from Newton's general definition of force. In other words, both these quantities give a 'true' measure of force, although the physical meaning of these two different measures was not yet clear. (Bernouilli had been the first of the Leibnizians to begin this kind of analysis, which, for those who know calculus, simply involves integrating Newton's force law with respect to distance or time, respectively; the Appendix gives the calculations.) Indeed, back in February 1738, Émilie had told Maupertuis, 'I agree with you that it is only a dispute over words', by which she meant that both 'living force' and 'quantity of motion' (momentum) existed. However, confusion over 'words' – and over the relationships between words like *vis viva* and momentum, formulae like $mv^2$ and $mv$, and processes like integration with respect to space and time – was the whole point: an adequate understanding of their physical implications would take another century.

*

In October 1742, with her husband's inheritance claim held up by one frustrating detail after another, Émilie finally told Argental she no longer 'flattered' herself that she could see it through, and she wanted to return to Cirey. (In the end, the marquis du Châtelet won his inheritance, but the negotiations involved selling the contested land and title at far less than he and Émilie had hoped for. Nevertheless, it brought them twice the amount they would have to pay out for their daughter's dowry and their son's first regiment. The aristocracy did not pay taxes – a contentious state of affairs – but they financed the king by buying promotions and positions. In the time-honoured monarchical tradition, commoners who were wealthy enough and useful enough could actually buy an aristocratic title.)

A few days later, Émilie wrote Argental again: 'We are finally here, my dear friend, at Cirey, which we love so much, and where I would like to spend my life.' But life at Cirey would never be the same again: although she did not yet fully realise it, Émilie's fairytale romance with Voltaire had not survived the lawsuit, which had taken up so much of her time in Brussels. Or maybe it was simply that the forty-eight year old Voltaire's health had become too poor to sustain sexual desire: this was certainly the reason he had given his friend Cideville, in July 1741, when he confessed that friendship rather than 'amour' had become the most important 'consolation' in his life, now that the vitality of youth had gone. Either way, it was he, not Émilie, who had fallen out of love, although he still loved her intensely as a friend, muse, intellectual companion and soul-mate.

In time, Émilie adjusted to the new relationship, in which she and Voltaire remained an inseparable, loving couple, although they were no longer lovers. They were both finding Cirey 'more charming than ever', and in many ways life went on as before. She still worried over him, and in return for her love and care, he promised not to return to Berlin, much to Frederick's anger. Voltaire would keep his promise, despite the fact that Euler had joined the Berlin Academy in 1741, and in 1746 Maupertuis would become its president. Over the years, the Prussian king would consistently reissue his invitation to Voltaire, who would repeatedly reply that Cirey was his court, and Émilie was his king.

Not that life was always easy at Cirey. In order to protect Voltaire from himself – or from uncomprehending visitors like Madame de Graffigny, who had copied out verses of Voltaire's poem on Joan of Arc and sent them to Devaux who, in turn, circulated them at court – Émilie had begun locking up any of Voltaire's works she thought might cause him trouble. For instance, Voltaire had continued to update and circulate the risqué political parodies in his 'farce on Joan', as he called his massive poem, and had recently sent some verses to Frederick, to help him 'relax after a hard day of government'. Émilie did not trust the young king to protect Voltaire's reputation by keeping the poem to himself; consequently, several months later when Voltaire sent more verses, he told Frederick he could only 'steal a few sheets' from the poem's 'severe guardian'. He complained that in order to send Frederick the whole poem (which ran to hundreds of pages), he would have had to 'use violence'. In fact, at times the intensity of Émilie and Voltaire's life together threatened to tear them apart: she was fiery, while he was prone to awful sulking moods. She could be imperious and controlling, not only sequestering his manuscripts but ordering him to change his jacket or insisting they did this or that together; he was egotistical, and he could be passively aggressive, deliberately hurting her by doing things without her. Sometimes he hurled awful words that made her cry for hours, and she occasionally threw at him a spoon or a plate. But always they made up. Usually, after a dreadful scene that had the servants tittering, one or other of them would begin cajoling, whispering secret words in English so the servants would not understand, and soon all would be forgiven and they would be laughing together. It was an intimacy very few people understood.

## *Chapter 10*

# THE RE-EMERGENCE OF MADAME NEWTON DU CHÂTELET

In 1744, Voltaire was reworking *Elements of Newton's Philosophy* for a new edition while Émilie was preparing a new edition of *Fundamentals of Physics* that included her reply to Mairan on *vis viva*. Voltaire told Condamine, 'Madame du Châtelet is clarifying Leibniz, which is very difficult, while I am muddling up Newton, which is very easy.' In April, Émilie wrote Argental that she and Voltaire were both finding Cirey enchanting, while Voltaire wrote his old school friend, the marquis d'Argenson (who was now a foreign minister in Louis XV's government): 'Cirey is charming, it is a jewel. Come here, monsieur – try to have business nearby. Madame du Châtelet loves you with all her heart, and desires your company as much as I do.' But another visitor came to Cirey instead, and he would change Émilie's life. He was the young Vatican-based French Franciscan priest and mathematician, Father François Jacquier, who, with his colleague Father Le Seur, had recently produced an annotated edition of the *Principia*.

Jacquier admired Émilie's *Fundamentals*, and had recently overseen its Italian translation (which Laura Bassi was now using with her students). He stayed at Cirey for two months over the summer of 1744, and it was during her conversations with him that Émilie conceived the idea of a French translation of the *Principia*. Jacquier's recent edition was in Latin, the language in which Newton had written, but Latin was losing its popularity as a universal language of scholarship. How much more accessible the theory of gravity would be to her compatriots if they could read it in their own language! Andrew Motte had published the first English translation of the *Principia* fifteen years earlier, but Émilie would be the first person outside Britain to translate what is still widely considered the most significant book in scientific history. Today there exist translations in all the major Western European languages, as well as many others, including Chinese, Japanese, Russian, Romanian and Mongolian.

Émilie also planned to add a scientific commentary to her translation, in which she would discuss new developments of Newton's work by later mathematicians. This would make her version the most up to date in the world in the mid-eighteenth century. She swore Jacquier to secrecy about her bold undertaking, and there is no mention of Newton in her letters to friends like Argental. However, Voltaire inadvertently dropped hints in his letters by calling her Madame Newton du Châtelet.

A year later, in June 1745, Voltaire would tell Algarotti, 'Émilie is still buried in the profound and sacred abysses of Newton.' It was an apt image for the mathematical terrain she had to master: not even Bernouilli could read the *Principia* easily. This difficulty was not only because of its challenging content, but also because Newton's book was initially put together rather rapidly, under Halley's urging. There was also the fact that Newton used geometric constructions rather than algebraic equations when he wanted to talk about geometric shapes like spheres, ellipses, flattened spheroids and the like. This led Euler, who was fast becoming one of the leading mathematicians in the world, to complain that many of Newton's proofs were so idiosyncratic that no-one could apply them to other similar problems.

As far as the basic theory of gravity was concerned, however, Newton advised his readers, even those who were mathematically adept, that to understand the theory (which is the subject of Book 3 of the *Principia*), they need only read the definitions, laws and first seventeen propositions of Book 1 (which has ninety-eight propositions in all), plus any other propositions from Books 1 and 2 that are specifically referred to in Book 3. Émilie would take his advice in structuring her commentary. Nevertheless, Newton's proof of the theory of gravity is an intricate web made of more than a dozen propositions and theorems. In some strands of the web, the proof of any one of these propositions might require the proofs of two or three other propositions, each of which may require proofs of yet other propositions, like nested Russian dolls. Émilie did an excellent job of sketching out the basic logical structure of Newton's theory, in Chapter 2 of her commentary, but she privately acknowledged it was 'terrible work' trying to follow Newton closely. And she had to follow him well enough to translate all 510 pages of the *Principia* from Latin into French.

She did it so well that modern historian Bernard Cohen regards Émilie's as one of the two 'great' eighteenth-century editions of the *Principia*, the other being that of Jacquier and Le Seur. In fact, Cohen utilised Émilie's interpretation of Newton in his new English translation of the *Principia* that was first published in 1999. In his preface, Cohen described the difficult process of translating passages in a technical text where the author's intention is not clear, especially a seventeenth-century book where certain terminology has since been modernised. He noted that he and his colleague Anne Whitman first translated Newton's final Latin version independently, and then they compared their rendering of some difficult passages with that of others: they found that 'the most useful works for such purpose were [twentieth-century Newton expert, Derek] Whiteside's translation of an early draft [of...] Book 1 of the *Principia*, and the French translation made in the mid-eighteenth century by the marquise du Châtelet'.

Back in November 1745, the arduous task of translation was still not completed, and Émilie wrote Jacquier of her slow progress: her work was being held up at the last minute by the fact that she was spending hours in the Minister of War's waiting room, trying to negotiate the purchase of her son's first regiment. Buying a regiment for Florent-Louis, who had just turned nineteen, meant he would be commander of his own infantry, although the process cost his mother many sleepless nights: she told Jacquier she was not getting to bed until four or five in the morning, but that when she had time, she was working on her translation – which, she cautioned, was still a secret.

In writing to Jacquier, Émilie also added news of her daughter. Jacquier had visited Gabrielle-Pauline at Émilie's request, and now she was pleased to tell him her daughter had just been awarded the prestigious post of lady-in-waiting to the Queen of Naples, a position 'she had strongly desired'. Back in August of the previous year, Voltaire had written Algarotti asking if he could use his influence to recommend Gabrielle-Pauline for this position. Voltaire had done this not only for Émilie, but also for Gabrielle-Pauline herself, with whom he still kept in occasional touch. For instance, in February 1746, he would write thanking her for her thoughtfulness in sending him 'a fine present of chocolate', adding that he wished he could drink some cups of her chocolate in Naples with her.

Émilie continued her letter to Jacquier: 'I will be delighted to be able to put at the head of the book, "Member of the Academy of Bologna".' This thrilling possibility referred to the fact that, after returning to Rome from Cirey, Jacquier had recommended Émilie's election to the Academy, one of the very few in the world that accepted women; no doubt Laura Bassi supported her nomination, as she had done for Voltaire the year before. (It seems Émilie and Laura never met, presumably because women did not travel much at that time: recall Émilie's husband would not allow her to visit England, and neither she nor Gabrielle-Pauline ever visited each other after the latter had moved to Italy, although they exchanged frequent letters.)

Émilie would have to wait until the late summer of 1746 for her official election to the Bologna Academy. In her thankyou letter to Jacquier, she said how important such recognition was for her as a woman, and that it would encourage 'persons of my sex . . . to engage in and cultivate the Sciences from which prejudice has so far appeared to exclude them'. Unfortunately, six months before she achieved this rare and well-deserved honour, her own work had been held up again: in early 1746, Clairaut told Jacquier he was reviewing Émilie's translation, on which, he said, she had 'worked like a slave all last year and part of this. She has the intention of adding to it a commentary, but it is not yet done, to tell you the truth.' He did not know the delay was caused by Émilie's irreparably broken heart.

She had been badly wounded when her relationship with Voltaire became asexual, but sometime during the winter of 1745–46 the wound was reopened by betrayal: Voltaire was in love with someone else. To complete the betrayal, he had not told her of his feelings, and she discovered the affair by accident: in protecting him from critical attacks on his reputation, she sometimes opened his mail so she could prepare him for any dramas – but this time she discovered, with growing horror, a love letter to the man she had believed loved her alone.

In this heartbreaking way she discovered that fifty-one year old Voltaire had recently begun an erotic correspondence with his niece, the widowed, thirty-three year old Marie-Louise Denis. (Émilie was thirty-nine at this time.) It seems to have been a rather strange relationship, not only because he still loved Émilie, but also because he would often complain to Denis of his ill-health – although he had recovered enough sexual desire to flirt and declare he wanted to spend his life with her. But he was so adept at flattering his correspondents it is difficult to unravel what he really felt. If actions speak louder than words, however, his decision to remain at Cirey with Émilie meant she was still the love of his life. And he would continue to be intensely jealous whenever he perceived an intimacy between her and another man, including Clairaut – although there is no other evidence of a romantic liaison between Émilie and her easygoing young mentor.

Nevertheless, she and Voltaire often clashed temperamentally, as I mentioned – she was as strong-willed and ambitious as him – whereas Denis seems to have been much simpler and easier to please. She relished playing his muse – and she even tried her hand at writing her own play, although nothing appears to have come of it. But at that stage her interaction with the notorious writer was mostly by letter, and she did not have to live day by day with his 'artistic' moods. By contrast, Émilie had often found herself in a 'cruel' situation, because 'Voltaire's charming sweetness changes under the worries and persecutions' that so often beset him. Denis would have experienced little of his dark side, and perhaps she offered him a reflected image of himself that was always happy.

Émilie was so devastated to learn she was not the only love in Voltaire's life that she put away her Newton. Unlike Maupertuis, she did not use mathematics as a calming distraction in times of emotional upheaval, because she could not concentrate on anything else when she was emotionally distraught. But she did leave a record of her feelings about the slow breakdown of her *grande passion*, in a manuscript called *Discours sur le bonheur* (*Discourse on Happiness*). She had already written on the importance to happiness of 'the love of study', but now she added that the passion of romantic love should not be excessive, or your lover will grow bored with the certainty of your love, and 'this is a fact of the human heart'.

Her recipe for happiness included not only moderating the passion of *amour* and cultivating a passion for study, but also avoiding prejudices, being healthy, and not dwelling on sad ideas (including 'never permitting our heart to conserve any taste for anyone who has lost his taste for us and ceased to love us'). It also included never feeling guilty ('never repent'), and being able to enjoy the pleasant things of life. She also noted that study was especially important for women: a love of study 'makes our happiness depend only on ourselves. Let us preserve ambition, and above all know what we want to be; let us decide on the route that we want to take to spend our life, and try to strew it with flowers.'

Today, we are used to reading such counsel in popular self-help books, but in an era when women were denied higher education and in whom ambition was considered shocking, the *Discourse on Happiness* was a revolutionary cry for female independence. Its advice is still relevant, and it has been republished in recent times, but the first edition was not published until after Émilie's son died in 1793: it seems Florent-Louis was extremely protective of his mother's memory, and he did not want her relationship with Voltaire laid bare. Perhaps he was also mindful of how it might reflect on him and his family's reputation – he was, by then, the duc du Châtelet. He was right to be concerned, because the eventual publication of the *Discourse* aroused snide comments about its author's turbulent and bohemian sex life.

By 1747, Émilie's love of life had reasserted itself sufficiently for her to return to her translation of the *Principia*. In February, Voltaire wrote to Algarotti referring proudly to 'our immortal Émilie' and her work on Newton. The War of the Austrian Succession was in full force, and Émilie had trouble getting letters through to Jacquier in Rome (and also to her daughter in Naples); but in April he received a letter in which she told him, 'I am still very occupied with my Newton. It is being printed now, and I am checking the proofs, which is very tedious, and I am working on the commentary, which is very difficult.'

The first part of Émilie's commentary would ultimately include an overview of the theory of gravity, including the logical structure of Newton's proof (the 'nested dolls' mentioned earlier), as well as relatively accessible verbal summaries of the more complex topics in the *Principia*. Her chapter on the shape of the earth also discussed Maupertuis's and Condamine's work, while the chapter on the tides included a discussion of recent papers updating Newton's work on this topic by Daniel Bernouilli, another son of the redoubtable Jean Bernouilli. There were also chapters on the precession of the equinoxes and on the theory of satellites and comets, including cutting-edge updates by Clairaut. There would be about 110 pages in this part of the commentary, which would be followed by about seventy pages of selected mathematical proofs.

In July 1747, Émilie wrote again to Jacquier, telling him her translation of Newton's Book 1 was nearly all printed, but that she was still working on the commentary, for which she was making extracts from Clairaut's work. Initially, historians would use this last statement against Émilie, claiming all her work was derivative of Voltaire or Clairaut, but the early chapters of the first part of her commentary were entirely her own work, and her struggle to gain mastery of this material is clearly evident in her letters to Maupertuis. For instance, nearly a decade earlier, she had written to Maupertuis with questions about the seemingly bizarre conclusions of Newton's 'superb' theorems; now, in her commentary, she was quite able to incorporate these into both her summary and her mathematical analysis. In the later chapters, she did draw heavily on Clairaut and Daniel Bernouilli, in order to describe their work. But her essay on fire, her reply to Mairan on *vis viva*, and her letters to Maupertuis about scientific subjects, all show that she was more than capable of distilling her mentors' work and making her own judgments on it. Furthermore, historian Judith Zinsser pointed out recently that Émilie's original notebooks prove her mathematical ability, because they show her work-in-progress on the calculations for the second, mathematical part of her commentary. (Incidentally, the manuscript of her unique version of *The Fable of the Bees* – which had remained unpublished, apart from being hand-copied and circulated within her own circle – and her annotations on some of Voltaire's manuscripts, were not discovered until the 1940s, when historian Ira O. Wade discovered them in St Petersburg. The discovery of *The Fable* revealed the independence of Émilie's early thinking – recall she had critiqued even Locke – while her annotations proved she had contributed to Voltaire's ideas and had not simply appropriated them.)

Émilie began her commentary by summarising the history of astronomy before Newton, noting in particular the pioneering work of Hooke and Kepler. Hooke was Newton's contemporary and bitter rival, and was famous for numerous discoveries and inventions – for instance, he was a pioneer in the science of microscopy, and Hooke's law of force for stretched springs is named in his honour. Émilie pointed out that more than a decade before the *Principia* was published, Hooke had written that every celestial body had its own 'attraction or gravitation towards its own centre'. He also suggested that the earth influences the sun and the moon, and vice versa, and he suspected gravity became weaker with distance, according to an inverse-square law. On the face of it, it is no wonder Hooke always claimed Newton had stolen his theory. Furthermore, Hooke's correspondence with Newton was the impetus for Newton's realisation that centripetal force should be directed inwards, rather than outwards as in centrifugal force.

Apart from this last important insight, however, Newton had had all the other pieces of the gravitational puzzle well in place at the time of his correspondence with Hooke – including a mathematical proof of the inverse-square law, which Hooke rashly accused him of plagiarising. This was despite the fact that Hooke himself publicly admitted he had not yet been able to prove any of his conjectures; the arguments he gave were 'hopelessly wrong', according to modern historians. Besides, others, notably Halley and his colleague Christopher Wren, had also suspected planetary motion might be subject to an inverse-square law, but they, too, had not been able to prove it. Émilie summed up the situation well when she wrote in her commentary that the work of Kepler and Hooke highlighted 'the distance between a truth that is suspected and a truth that is proven, and it also shows how little service to science even the greatest minds provide when they are not guided by mathematics'. Newton, by contrast, had used 'the most profound mathematics to deduce the law of gravity, and the principle suspected by Kepler and Hooke became, in his hands, a fecund source of admirable and unexpected truths'. (Incidentally, in the wake of Hooke's accusations, Halley once again had had to come to the rescue of the *Principia*. Newton never seemed to outgrow the insecurity of his abandoned childhood (his father died before he was born, and, on her remarriage, his mother left three-year-old Isaac behind with his grandparents); at any rate, he was prone to throwing tantrums, and when Hooke claimed the inverse-square law as his own, Newton refused to continue work on the *Principia*. Eventually, the ever-patient Halley managed to persuade him to go on. In his preface to the *Principia*, Newton gratefully acknowledged Halley's encouragement, including the fact that it was Halley – 'a man of the greatest intelligence and universal learning' – 'who started me off on the road to this publication'.)

After her historical overview, Émilie gave a detailed account of Newton's proof of the theory of gravity. As I mentioned in Chapter 2, Newton based his theory on the fact that the moon's motion about the earth can be explained by assuming it is caused by the earth's gravity, and in the Appendix I have given a simplified overview of his proof of this. (My version is much simpler than the one in Émilie's commentary but it is similar to the one she gave in her *Fundamentals*;

the *Elements* contained an even simpler sketch.) Newton then went on to show that gravity is a universal phenomenon, applying to all material bodies throughout the universe. But, as Émilie showed in detail in her commentary, it was not a simple case of proving the theory of gravity for the moon's motion and then *assuming* it must also work for planetary motion. Newton's proof that the planets orbit the sun in gravitationally induced elliptical orbits combined the proofs of all those nested propositions and theorems, including not only the inverse-square law for elliptical motion, but also the planetary versions of Kepler's second and third laws – along with Newton's derivation of the law of equal areas for centripetal force – together with some of his so-called 'superb' theorems. The whole conception was utterly marvellous, and it was based on an astonishing chain of mathematical reasoning that dazzled his contemporaries, even those who rejected the theory of gravity itself.

The second part of Émilie's commentary recast Newton's geometrical proofs of key propositions in terms of Leibnizian calculus – including proofs of some of the 'superb' theorems. Recall one of these theorems proves that the earth's gravitational force on another body located on or beyond its surface is calculated from the inverse-square law, where the distance is measured from the centre of the earth. This means that in calculating the earth's gravitational force, it is as though all the matter in the earth is concentrated at a single point at the earth's centre. This is a remarkable result – which, as Émilie pointed out, applies to *any* spherical object – and Newton proved it in the fourth 'superb' theorem, Proposition 74.

Fundamental to Newton's geometrical proof of this proposition is the intuitive 'limit' concept of calculus: he proved his argument by mathematically decomposing the matter in a solid sphere into layers of infinitesimally thin spherical shells. He then showed that the sum of the gravitational forces emanating from all the shells gives the inverse-square law, with the distance calculated from the centre of the sphere. Émilie used the same idea, but she described the shells in terms of algebraic equations rather than Newton's geometrical diagrams, which meant calculus algorithms could be used where necessary. ('Integral calculus' is the process of adding an infinite number of infinitesimal quantities, and 'adding' the tiny gravitational forces from the infinitesimal shells of the sphere is more systematic when done with algebraic algorithms. After all, few people have Newton's idiosyncratic geometrical insight.)

In another of the 'superb' theorems (Proposition 73), Newton showed that the gravitational force on each particle *inside* a solid sphere is *not* an inverse-square force. Rather, it is a force directly proportional to the particle's distance from the centre. This surprising result is a consequence of Newton's proof that for any two spherical particles of matter, the inverse-square law holds, but that when many particles are acting (such as all the particles of which the earth itself is composed), there may be a different averaged-out law, depending on the distribution and shape of the matter. (Recall that in calculating the amount of flattening of the earth, Newton used an analogous theorem – Proposition 91, corollary 3, which is for spheroids rather than spheres – to find the required length of his imaginary columns of fluid inside the earth.) It is easiest to see how this works – how the inverse-square law is averaged out when many particles are interacting – by looking at yet another of Newton's theorems, the first 'superb' theorem, Proposition 70. This intriguing theorem says that for a particle placed anywhere inside a *hollow* sphere (that is, inside a spherical shell), there is *no* net gravitational pull from the shell. In the Appendix I have given a sketch of this proof, using Newton's geometrical way of thinking rather than calculus algorithms.

Late in 1747, however, Émilie's project was held up by a devastating announcement from Clairaut to the Parisian Academy of Sciences: the inverse-square law of gravity was incorrect. At least, it did not give the correct description of the observed orbit of the moon, which was *not* exactly circular, as Newton himself realised.

According to Newton's theory, the moon would orbit the earth in a perfect circle or ellipse if it were affected only by the earth's gravity (circles and ellipses being the only possible shapes for closed orbital motion caused by an inverse-square force). Working out the theoretical gravitational attraction between the moon and the earth, or between a planet and the sun, is a relatively simple 'two-body' problem, but in reality the moon is also significantly affected by the sun's gravity, so its orbit needs to be solved as a 'three-body problem'. Taking account of the interacting gravitational forces between these three bodies makes the problem completely different mathematically from the 'two-body' case, and much harder to solve. This is because the gravitational influence of the sun keeps changing as the moon travels around the earth – because the moon's position with respect to the sun changes during its orbit of the earth, so that it becomes sometimes closer and sometimes further from the sun. This means the sun's gravity distorts or 'perturbs' the moon's orbit around the earth by different amounts depending on the distance between the moon and the sun at any point. Newton pioneered what we now call mathematical 'perturbation methods' to deal with this problem, which can only be solved approximately, so there are no 'exact' solutions like those giving ellipses and circles for two bodies.

In trying to extend Newton's work on the lunar orbit, Clairaut had finally concluded the only way to make the theory of gravity consistent with observation in this case was to accept that Newton's inverse-square law was only an approximation; the true law, Clairaut suggested, contained *two* terms: an inverse-square and an inverse-fourth power.

His colleagues D'Alembert and Euler were also working on the problem of the moon, and they, too, considered tampering with Newton's theory: D'Alembert suggested there was another kind of force acting as well as gravity, while Euler advocated bringing back vortices. This kind of sceptical checking and rechecking is the proper work of mathematical physicists: none of the great theories of physics is the last word on physical reality – because a mathematical theory can never be more than an idealised approximation of the real world – and so all theories can be improved in order to more closely fit with physical observations. But Euler and D'Alembert were more cautious than Clairaut about rushing into print with their suggestions: D'Alembert told a colleague, 'I plan to publish next year . . . but I am so much afraid of making assertions on such an important matter that

I am in no hurry to publish anything on the subject. Besides, I will be very sorry to overthrow Newton.'

Clairaut, on the other hand, took Newton to task, criticising him for being so secretive about his method of calculus, and for giving 'too few words to explain his principles' in the difficult parts of the *Principia*. Newton had said the moon problem 'causeth my head to ache', but clearly his incomplete work on the subject was now causing Clairaut much agony!

Over the next year, Clairaut would persevere with his calculations, and after a prodigious amount of effort, he managed to take the analysis of the lunar orbit to a new level, through the refinement of Newton's 'perturbation methods'. He discovered that the inverse-square law correctly gives the orbit of one body about another, but that a fourth power term must be added in order to find an orbit that is perturbed by the gravity of additional bodies. In other words, Clairaut's inverse fourth power approximated the effects of the other perturbations, but it was not part of the law of gravity itself. Consequently, he would be forced to retract his claim against the inverse-square law. But, as Euler pointed out, Clairaut's reworked calculations provided the first irrefutable proof of the validity of Newton's theory.

Back in 1747, however, Émilie had had no option but to stop the printing of her work on the *Principia*. With her work held up, and with her heart still hurting after her discovery of Voltaire's infidelity, she threw herself into family business again. In considering the careers of her husband and son, she had already begun to focus her attention on the possibilities at King Stanislas's court at Lunéville, in Lorraine, the marquis du Châtelet's ancestral territory. A decade earlier, Stanislas had abdicated the throne of Poland after many years in exile, having been deposed by Augustus II in an alliance with Peter the Great of Russia; he reigned over Lorraine (as a duke or regional leader) by arrangement with the French king, his son-in-law.

Shortly after Clairaut's dramatic declaration that the inverse-square law was wrong, Stanislas invited Émilie and Voltaire to set up rooms at the Lunéville court; the change of scene was no doubt particularly welcome for Émilie, since the ghosts of happier times still haunted Cirey. Not surprisingly, it appears she did not consult much with Voltaire on her plans: after they had set up their new residences, he told Argental he was not sure whether she planned to stay in Lunéville for the rest of the month, but, whatever she decided, he was 'only a small planet in her vortex', and he would 'hobble along in her orbit'.

On the same day, he wrote Madame Denis, 'My dear, here I am, your wandering friend, in Lunéville, with a King who has nothing of the king about him except generosity and a greatness of soul. But I would infinitely prefer your boudoir to all the courts.' After describing the delightful atmosphere at Lunéville, he declared he could only be happy if she was there too, 'but destiny still separates us.' However, destiny was about to intervene.

Stanislas's relaxed and cultured court brought out Émilie's remarkable vivacity. She still adored dressing up with her 'pompoms' and diamonds, and she shone in the regular theatrical entertainments, just as she had done at Cirey. She played comedy brilliantly, and her voice was admired so much that in those early weeks at Lunéville, in February 1748, she gave three performances in a row as the operatic lead in *Issé*. At forty-one, a relatively advanced middle age in those days, she was so radiant and charming that she caught the eye of the thirty-one year old soldier-poet, Jean-François, marquis de Saint-Lambert.

Saint-Lambert was a friend of Madame de Graffigny and her young Lunéville-based confidant, Devaux; he was also the sometime lover of the extraordinary marquise de Boufflers, the self-styled 'lady of sensual delights'. The beautiful, intelligent Cathérine de Boufflers was also the favourite mistress and unofficial consort of the recently widowed King Stanislas. She and Émilie would soon become friends, but she and Saint-Lambert had been friends since childhood. They were still occasional lovers, but, unlike Émilie, Boufflers was a coquette who used her wiles to flirt with all the men at court.

Émilie was attracted by Saint-Lambert's poetic sensibility, and also, no doubt, by his attentions. In February and March, they flirted deliciously, playing the game of love by having their maids deliver little love notes to one another, indulging in secret meetings, and enjoying the delights of romance without yet becoming sexually intimate. Saint-Lambert assumed the relationship would remain this way – just another game of secret, superficial seduction – but that was not Émilie's style.

## *Chapter 11*

# LOVE LETTERS TO SAINT-LAMBERT

Émilie's letters to Saint-Lambert are proof of her forceful, achingly intense way of loving. Even the little notes she sent in those first weeks at Lunéville are intense: 'You are not made for *my amusement*, but for my happiness, and for my whole occupation if you want.' But then followed another note, written at two in the morning: 'You treated me so coldly today, you seemed so little occupied with me . . .' Perhaps she had misread his intentions, perhaps she had imagined too much. She knew love was risky with a man like Saint-Lambert, who was still young, and who was a favourite with the women at court.

The main correspondence between them began in late April 1748. It is often painful to read Émilie's passionate, compulsive letters, since they were mostly written during their 'awful separations'. Sometimes she wrote three or four letters a day, telling Saint-Lambert how much she loved him, and how she would try to be worthy of his love, and how she was sorry if she had upset him with her intensity or her directness. She knew it was risky to want so much from him, but 'when my heart leads [my head], it has no common sense'.

Saint-Lambert, on the other hand, remained 'half tender, half detached', as Émilie put it. But despite her doubts about him, he was capable of making her deliriously happy: her letters testify to the fact that they shared some blissful times together, especially in the summer of 1748, when he was free of military service for a while.

'The tears of lovers appease the gods,' he wrote in a poem for her. 'Louis calms the earth, he gives me back to myself. I no longer sell my time to the quarrels of kings.' She replied, 'You are the only one who has made me feel that [my heart] is still capable of loving', but 'I cannot be happy loving you if you do not love me to excess . . . In making love, enough is never enough.' Nevertheless, she wanted him to take more interest in her life, especially her resumed work on the *Principia*, which she referred to as 'my Newton': 'It is a project that is very serious and very essential for me.' She often had to remind him she could not always get away to see him, because: 'This book is awaited, promised, begun two years ago, my reputation depends on it.'

Unfortunately, snide critics had a different view of her reputation: several months earlier, Voltaire had written a letter defending her against an anonymous claim that 'it is a madwoman who knows atoms better than her own family, and who believes herself a philosopher despite the disarray of her passions, through which she has a confused knowledge of useless things'. Since Émilie supported Leibniz's proof that atoms did not exist, this nasty gossiper had not even read her work! Voltaire was outraged at these 'odious and unjust claims against a woman this author does not even know'. He added that it was 'a strange way to speak about a woman of the first order who is respected in Europe for her unique talents, and who – far from abandoning her family – has procured important positions for her husband and son, and has married her daughter to a first lord of Naples, and who has worked to liquidate all the debts of her estate'. It was not the first time Voltaire had defended Émilie against malicious attacks occasioned solely by the fact that she was a woman who dared also to be a scholar. But she had also been attacked for daring to be a passionate woman openly in love with Voltaire – and soon the gossips would be sniffing out the scandal of her growing passion for Saint-Lambert.

For many months Émilie did not tell Voltaire about her new love. After all, he had deceived her with Madame Denis, and she assumed their private lives were now separate. Nevertheless, she tried to be discreet, for his sake and her husband's. When Voltaire finally learned of the affair, he was outraged. Émilie was stunned at his anger, given there was no 'illusion of attraction and love' between them, as she put it, and she later reassured Saint-Lambert, 'There is nothing to do with such a character but avoid him and blush at having loved him.'

It seems Voltaire's rage was prompted largely by fear for his reputation: because his and Émilie's relationship was now widely accepted as an unofficial marriage, it was he who had been 'cuckolded', not the marquis du Châtelet. Nevertheless, he might well have been sincerely jealous on discovering Émilie's love for Saint-Lambert, because he knew first-hand that she would see this new grand passion as a chance to make something truly magnificent. Only a few months earlier, he had written a poem about two jealous gods, Mind and Heart (*Esprit* and *Amour*): on one single occasion, these gods momentarily reconciled – and produced Émilie!

Denis herself had noticed this special combination in Émilie: ten years earlier, during a visit to Cirey, she had written, 'There is no passage from the best philosophers that she does not recite to Voltaire. A woman of great intellect, she is also very pretty and uses every art to seduce him. He is more captivated than ever.' Even now, despite his declarations of undying love in his letters to Denis, and the occasional erotic anticipation of a rendezvous with her, he must have known he would never again experience the kind of passion he had known with Émilie. In fact, while Denis relished the possibility that Voltaire might finally be free of 'that woman', as she now called Émilie, Voltaire could not so easily break the tie between them. Once he recovered from his shock and calmed down over Saint-Lambert, the unique bond between Voltaire and Émilie reasserted itself: once again they would often find themselves talking long into the night, enthusiastically discussing their ideas as they had always done.

In late January 1749, however, Émilie discovered she was pregnant, at the age of forty-two. In aristocratic society, to be pregnant at such a late age was humiliating enough, but to be carrying the child of her lover rather than her husband was a social crime that could have led to Émilie's banishment to a nunnery. Years earlier, Émilie had written a moving version of the section from Mandeville's *The Fable of the Bees* on the unhappy situation of women who had children outside marriage. To avoid such a fate, Madame de Tencin, the famous salonnière, had abandoned her 'illegitimate' baby on the steps of a Parisian church, St Jean Baptiste Le Rond; the foundling grew up to be Jean Le Rond d'Alembert, the celebrated mathematician and colleague of Clairaut, and future mentor of Laplace.

The marquis du Châtelet did not disown his wife, but stood by her as he had stood by her with Voltaire. When the pregnancy became public, everyone assumed the baby was Saint-Lambert's, and Émilie felt 'a thousand dagger blows' at the thought that she had exposed her good-natured husband to public ridicule.

Insensitive to her plight, Saint-Lambert continued his on-off way of relating to Émilie, even to the extent of resuming a dalliance with his former lover, Madame de Boufflers. Consequently, if Émilie's letters to Saint-Lambert were intense in the first year of their relationship, those written when she was pregnant and alone are truly heart-rending. Terrified she would lose him to a younger woman, ridiculed by society gossips for being pregnant at such an age – and scorned by her twenty-one year old son, who was afraid of losing some of his inheritance to this unexpected unborn rival – she poured all her demons into her letters to her wayward lover: '[I fear] you will leave me, without reason . . . you will abandon me in my state, you will make me die of grief . . . you will betray me for someone [who does not deserve you]. You will sacrifice me in favour of her! There is no longer any truth, I no longer believe even in the truths of mathematics! I only want to believe in the truth of your heart . . .'

A few months later, however, she had accepted her situation and was once again working on her Newton. Clairaut was so impressed with her progress that he spent a considerable amount of time with her, checking her calculations and reviewing her commentary. Now that he had resolved the problem of the moon's orbit so that the inverse-square law did not need to be modified, his encouragement gave her renewed enthusiasm for the challenge of completing her project. And she certainly needed such encouragement: although she had a phenomenally quick mind – Voltaire said she could divide numbers of up to nine digits in her head, to the amazement of established mathematicians – she found working on her commentary extremely arduous. But she did not give up, and in May 1749 she once again had to reassure Saint-Lambert that she adored him despite her mathematical preoccupation: 'Do not reproach me for my Newton, I am punished enough – I have never made such a sacrifice for reason as to stay here and finish [this book]. It is an awful job, for which one needs a head and a constitution of iron.'

The popular perception of doing mathematics is that it is difficult but singularly satisfying, because it is so logical and so certain, and this is, indeed, one of the reasons mathematicians are drawn to their field. It is utterly exhilarating to follow through a complex chain of reasoning and come to an absolutely certain conclusion that can be summed up in a single sentence, or to prove a complex statement in a few elegant, unassailable lines. But the path to such an achievement is often excruciatingly delicate, as Clairaut had discovered with his hasty pronouncement on the inverse-square law. Solving new mathematical problems often involves pages and pages of calculations, each line of which must be absolutely correct. Something as simple as a missed minus sign thirty pages back can render the whole lot useless, so that days of tedious work were all for nothing. Worse, the thrill of apparent success can be ruthlessly ripped away when one cautiously, meticulously, with beating heart, checks and rechecks an exciting conclusion, only to discover a tiny mistake buried deep within those arcane hieroglyphics. Émilie's statement about needing a head and a constitution of iron is exactly right!

But sometimes Émilie feared she was not destined to see the fruit of her laborious work. She had spent much time in her *Fundamentals of Physics* trying to resolve Leibniz's ideas on free will within a Newtonian framework of deterministic natural law, but now she could not shake off a premonition that she was doomed to die in childbirth, like so many women. With her deathly foreboding, she worked day and night on her commentary, sleeping only three or four hours a day, desperate to leave something of herself for the future. It was such gruelling work that she told Saint-Lambert she no longer loved Newton, she loved only him – although perhaps she was reassuring her lover as she wished he would reassure her: she told him, in late May, that she was working rather than socialising, and that it really was necessary that she stay put and finish her book before the baby was born, or else she would 'lose the fruit of my work if I die in childbirth'. But she quickly reassured him that she was marvellously well – she was being abstemious and looking after herself – and that the baby was moving a lot and she hoped it was as well as she was.

In the summer of 1749 – in the belief that Voltaire would no longer stand by Émilie now she was pregnant to Saint-Lambert – King Frederick renewed his efforts to entice Voltaire to his vibrant Berlin Academy. Voltaire replied that he would postpone his decision until after Émilie had had her baby, but Frederick responded tartly that surely she could have the baby without him, because he was not a midwife! For years Voltaire had flattered the king obsequiously, but now he replied with anger: no-one, he said, 'not even Frederick the Great who makes [everyone] tremble, can at present prevent me from fulfilling a duty that I believe is very indispensable. I am neither the maker of babies, nor doctor, nor midwife, but I will not leave, even for Your Majesty, a woman who might die in September. Childbirth seems very dangerous, but if all goes well, I promise you, Sire, I will come to pay court to you in the month of September.'

By late summer, Émilie decided she had done enough work on her commentary that she could stop her gruelling routine, for which she had moved to Paris, in order to check the printer's proofs. She and Voltaire settled happily back into Lunéville – he was working on his new play, *Catalina*, and she was putting the finishing touches to her Newton. When Saint-Lambert was able to get away from his duties, he joined them there, making Émilie blissfully happy.

In late August, Saint-Lambert had to go to Nancy on duty, and even Voltaire had become a little impatient; on 31 August, he told Frederick, 'Madame du Châtelet has not yet delivered. She has more difficulty bringing into the world a baby than a book.' Indeed, Émilie had finally finished her Newton: with Voltaire's encouragement, she was about to send the manuscript to the royal librarian, for safekeeping should she die, and as proof of her authorship.

While Voltaire was writing boys' talk to Frederick about how difficult it was to give birth to literary works like *Catalina*, Émilie wrote an anguished, late-pregnancy letter to Saint-Lambert: 'My belly has dropped terribly, I have such a pain in my back, I am so sad this evening that I won't be surprised if I have the baby tonight . . . I will be able to bear my sorrow more patiently when I know that you are [here] . . . I had hoped to work during your absence, but I am not able to. I am too sick, I have an unbearable backache, and a discouragement in my mind and in all of me except my heart.' She went on to apologise for not being able to tell him more about how much she loved him, but she didn't have the strength to keep writing.

Finally, on 4 September 1749, Émilie and Saint-Lambert's daughter arrived. It had all gone so well, Voltaire told a friend, that although the mother was now sleeping, she was not as tired as he was, having recently delivered his *Catalina*.

Over the next few days, Émilie made some final changes to the proofs of her *Principia*; the baby, named Stanislas-Adéläide, was with a wet nurse as was customary, and apart from the late summer heat, all seemed to be going well. But suddenly, on 10 September, Émilie took a turn. The king's doctor was called, and he was so alarmed he called in two more doctors for consultation; the patient was given drugs, which seemed to calm her, and nearly everyone went down to supper. Saint-Lambert and Mlle de Thil stayed with her, along with her chambermaid, and the faithful Longchamps, her former assistant, who was now Voltaire's valet. After a few minutes, Émilie suddenly groaned and gasped for breath as if she were suffocating.

It must have been terrible to hear those awful sounds as she desperately struggled to hold on to life. How much worse to see her fall back in silence . . . When the others arrived, in tears, they found a ghostly

Saint-Lambert paralysed with shock. The marquis du Châtelet was so upset he could not stand up, while Voltaire sobbed uncontrollably. A little later, he railed at Saint-Lambert like a madman, accusing him of killing his beloved Émilie.

*Chapter 12*

# MOURNING ÉMILIE

Émilie was buried at Lunéville with all the honours Stanislas could provide. Saint-Lambert immersed himself in an agony of private grief, and he would take over a year to recover from the nervous breakdown that followed his lover's death. 'I would never have believed him capable of such passion,' said Madame de Graffigny.

Meanwhile, broken-hearted Voltaire wrote letter after letter telling his friends that Émilie's death had left him 'without consolation on Earth'. He felt his miserable life was over, and he was going to Cirey to help sort out her papers, but 'it will be awful to go to see the house we had embellished so much, and where I expected to die in her arms . . . I have not lost a mistress, I have lost half of myself, a soul for whom mine was made.' While at Cirey, he began to re-read 'the immense metaphysical materials that Madame du Châtelet had assembled with a patience and a wisdom that frightens me . . . Ah my dear friend [Argental], they do not know what they have lost.'

Voltaire was one of the greatest writers of the century, and for many weeks after her death he continued to write poignant, powerful letters in tribute to Émilie. In mid-October, he thanked Condamine for his condolences, adding words that would have appalled Madame Denis: 'I have lost the support for my unhappy and lingering life. I have arrived from Lorraine overwhelmed with despair and illness. How is it possible that it is she who has died before me! . . . It is necessary to suffer, and to see suffering, to see death and to die. That is our lot.'

But he asked Condamine to tell Clairaut that he needed to talk to him, presumably about Émilie's scientific legacy, because he told another colleague, 'I hope you will soon see her Newton. She has done what the Academy of Sciences should have done. Any thinking person will honour her memory, and I will spend my life mourning her.'

Needless to say, I was profoundly moved when, two and a half centuries later, I held Émilie's handwritten *Principia* in my own hands. How much joy and suffering had gone into those pages – nearly a thousand of them, double-sided, arranged in three volumes, each page carefully numbered. I cannot bear to think how difficult this work must have been for her. You can see just a hint of the enormity of the task in the numerous crossings out, the inkblots, and the revisions made and stuck over the top of some passages, and the deleted passages literally cut out of the page. You can see a hint of her aristocratic circumstances, too, in the traces of gold edging on the paper. But you cannot see the pain, the ambition, the hopes she had had for 'her Newton' in those dark, lonely days when she feared she would not survive childbirth.

During Voltaire's intense period of mourning, the sardonic Frederick had noted, 'Voltaire declaims his affliction too much, which makes me think he will soon be consoled.' Frederick was right in that Voltaire's life did go on: plays had to be finished and performed, and gradually his letters spoke of things other than his grief. In January 1750, four months after Émilie's death, Madame Denis moved in with him; six months later, he moved to Berlin, while she preferred to stay in Paris. In Berlin, Voltaire wrote a moving eulogy for Émilie, which was to be included in the preface to the published version of her Newton in which he also declared: 'Here we witness two marvels: one, that Newton wrote this Work; the other, that a Woman translated and elucidated it.' But soon he moved on, and the proofs of Émilie's Newton lay forgotten. A year later, her baby died too, and was buried with her in Stanislas's churchyard.

Voltaire never returned to science after Émilie died; in fact, his interest had begun to wane after his inconclusive experiments on fire, although in the early 1740s he had made valiant efforts to understand the complex issues at the heart of the *vis viva* debate. Nevertheless, he continued to maintain his personal support for Newton's theory, although it was Clairaut who finally organised for Émilie's complete French version of the *Principia* to be published, in 1759. (An incomplete edition had been published a few years earlier.) He was motivated by an auspicious occasion. The year 1759 was not only the tenth anniversary of Émilie's death, but it was also the year of the first definitive return of Halley's comet.

In the *Principia*, Newton had proved that, theoretically, comets that stay in the solar system should travel in elliptical orbits, and he had devised methods for determining such an elliptical path from several astronomical sightings of a comet's position. Halley's achievement was to provide dramatic experimental confirmation of this theory: he identified a particular series of comet sightings, recorded over a period of two thousand years, as instances of the same comet (which became known as Halley's comet). With this data, he had worked out the size of the elliptical orbit, and predicted the comet would next return to our area of the solar system in 1758. However, in their journey around the sun or their parent planet, smaller celestial objects like comets and moons are particularly susceptible to gravitational forces from other large planets, which distort or perturb their orbits from the perfect ellipses or circles they would otherwise describe. Clairaut's work on the lunar orbit had enabled him to improve Newton's 'perturbation methods', and, over the winter of 1758–59, he had applied his new techniques to obtain a more precise prediction for the return of Halley's comet: Clairaut calculated that it would return, to the point on its orbit that is closest to the sun, in mid-April 1759 (give or take a few weeks). Émilie had noted in her commentary that if this comet did return as Halley had predicted, it would be 'a most flattering moment for the partisans of Newton'. How thrilled she would have been when her dear friend – her 'little Clairaut' – predicted its return to within a month of its actual appearance in March 1759. It was, indeed, a dramatic confirmation of the power of Newton's mathematical theory of gravity to make accurate astronomical predictions.

Nevertheless, the publication of Émilie's *Principia* did not immediately fulfil her hope that her work would live forever. Instead, her intellectual contribution remained marginalised for two hundred years, considered derivative of the great men who had supported her. Wade's discovery of her manuscripts in the 1940s was the beginning of a re-evaluation of her legacy, but it was not until the 1970s that biographers began to take Émilie seriously. Before then, according to historian Elisabeth Badinter, she was simply 'too free, too intelligent and too ambitious' to be a suitable role model for young women.

In the last forty years, she has become a heroine not only for young women, but also for 'any thinking person', as Voltaire predicted. Her *Principia* is still the only complete French translation, and her other writings show her extraordinary courage, honesty, passion, wit and charisma. She broke so many stereotypes about women *and* about mathematicians, stereotypes that lingered until the eve of the twenty-first century. In particular, she showed it is possible to be both emotional and rational, both intellectual and sexy. She was truly the 'divine Émilie'.

As for Voltaire, Frederick's court had not been the literary sanctuary he had imagined it would be, with its petty intrigues and power plays, and in the end his own penchant for trouble saw him fall out with both Maupertuis (the Berlin Academy's president) and Frederick himself. Consequently, in 1755, he moved to Geneva, eventually buying a nearby château at Ferney, just inside the French border. Madame Denis joined him there and became known as the 'lady of Ferney'.

In the early 1760s, Voltaire led a successful fight to clear the name of Jean Calas, a Protestant who, in 1762, was condemned to a gruesome death on trumped-up charges by Catholic zealots in Toulouse. As a result of Voltaire's campaign, Louis XV paid Calas's widow a considerable sum in compensation, although no charges were laid against Calas's murderers. Voltaire later worked on behalf of other victims of religious fanaticism; his good work on behalf of such victims – and the infrastructure and support he provided for the town and its people – led to his revered nickname, 'the patriarch of Ferney'.

Walking around Paris today, it is hard not to like and admire Voltaire, whose kindly face smiles down from the many statues erected to this great poet, dramatist, essayist and revolutionary hero. His wit and courage in writing on freedom and tolerance had made him a legend in his own lifetime – so much so that in 1778, the eighty-three year old patriarch of Ferney returned to Paris for a triumphant opening night of his tragedy, *Irène*. After decades in exile and a lifetime of harassment, the excitement of such a hero's welcome was overwhelming, and he died shortly afterwards.

More than a decade later – in 1791, just two years after the French Revolution – crowds followed Voltaire's cortege for more than eight hours as his remains were transferred to the Panthéon, that beautifully domed former church that was now a revolutionary temple not to God, king or aristocratic privilege, but to meritocracy and the great men of France. (The first woman to be so honoured, on the strength of her own merit, was Marie Curie, whose ashes were reburied in the Panthéon in 1995.)

Two years later, the early idealism of the Revolution was stifled by inherited problems of longstanding government debt and by foreign wars and the threat of invasion. In the wake of such problems, faction fighting broke out among the leaders of the Revolution, most of whom were completely unprepared for the difficult task of government in such tumultuous times. By the summer of 1793, the extreme left wing of the 'Jacobin' faction had gained control of the National Convention, led by the notorious Maximilien Robespierre, and by October, the 'Reign of Terror' had begun.

Ideologically, the Jacobins were inspired by Jean-Jacques Rousseau's egalitarian version of Hobbes's authoritarian social contract philosophy, in which the people cede their sovereignty to the state, which pledged itself to oversee the equal distribution of wealth among its people. Rousseau had spelled out his political theory in 1762, when he published his *Social Contract*, which opens with the immortal line, 'Man is born free; and everywhere he is in chains.' To maintain political stability, Rousseau believed the individual's rights were subordinate to the 'general will' of the collective whole. In a travesty of this naively idealistic philosophy – and in fulfilment of Locke's earlier warnings about the dangers of absolute government – Robespierre's faction used the authority of Rousseau to support their view that anyone who opposed the 'collective will', as determined by the elected government, was a traitor. Consequently, thousands of people were executed during the ten-month frenzy that became known simply as 'The Terror', from October 1793 until July 1794.

Parisian playwright Jean-François Ducis left an eloquent eyewitness account of those dreadful days, in a letter to a friend: 'What are you talking about, asking if I am occupying myself writing tragedies? Tragedy is running in the streets. If I set foot outside, I have blood up to my ankles . . . It is a harsh drama, this, where the people play the tyrant . . . Believe me, I would give half of the rest of my life to be able to spend the other half in some corner of the world where liberty was not a bloody fury.'

In the midst of this bloody fury, Émilie's son, Florent-Louis, fell victim to the guillotine. He had been a colonel in Louis XVI's army – a rather inept and unpopular colonel, it seems – and he was executed on 13 December 1793, aged sixty-six. He had no children, and his sister, Gabrielle-Pauline, had no surviving children, so with his death, Émilie's direct line was extinguished. And when the Terrorist zeal infected the patriots of Lunéville, they ransacked the graves in Stanislas's churchyard. Madame de Graffigny's young companion, Devaux – by then an old man of eighty-three – was the only survivor of the dazzling days at Stanislas's court: he witnessed a shocking desecration of Émilie's grave, in which her bones were scattered and her jewellery and finery mocked and stolen by uncomprehending 'citizens' of the new republic.

When the mob had gone, Devaux lovingly replaced Émilie's remains in her grave. There was no inscription on her black, marble tombstone, but the old man regularly kept a silent vigil in honour of her memory, sitting by her grave and remembering the glory days of the *philosophes* – the days of hope, through faith in reason, before reason temporarily turned into madness.

*Chapter 13*

# MARY FAIRFAX SOMERVILLE

The Terror notwithstanding, by the end of the eighteenth century much had changed in France in a way that would have pleased Émilie and Voltaire. Politically, the autocratic monarchy had been abolished and the church's power over secular life curtailed. Scientifically, Newton reigned throughout Europe, and it was French mathematicians who had done the most to ensure such an outcome. Building on the pioneering work of Émilie's friends – notably Maupertuis, Clairaut and D'Alembert, and their Swiss colleagues, Euler and the Bernouillis – the new generation of French mathematical physicists were not only firmly Newtonian, but mathematically they had also eclipsed Newton's British disciples. In particular, Pierre-Simon de Laplace had significantly improved the Newtonian theory of the tides, and he had also begun to tackle a conundrum left hanging by Newton – a conundrum that had led Leibniz, Émilie and others to misunderstand Newton's position on *vis viva*.

Laplace's work is the mathematical foundation of Mary Somerville's story. However, to appreciate the obstacles she faced in becoming a mathematician in the first place, it is important to note that not much had changed for women since Émilie's time. In terms of mathematical achievements, the most famous of Émilie's female contemporaries had been the Italian Newtonians, Laura Bassi and the child prodigy Maria Agnesi, who, like Bassi, held an honorary professorship at Bologna. In 1748, with the help of the mathematician Vincenzo

Riccati, Agnesi had published a textbook on calculus (whose title in English is *Analytical Institutions for Italian Youth*). But she was not interested in a mathematical career, which seems to have been her father's idea: after he died in 1752, she gave up mathematics altogether, at the age of thirty-four. She refused to marry (as the eldest of twenty-one children, she had no doubt had enough of caring for babies), and devoted the rest of her life to charitable service, beginning with the education of her youngest siblings.

By contrast, Laura Bassi was ambitious. She never stopped fighting for her rightful place as a fully recognised academic – she had held honorary or semi-professional positions for decades. Her tenacity, and the support of her husband, eventually paid off: in 1776, when she was sixty-five years old, she was finally granted a full professorial position in experimental physics. She thus became the world's first 'real' female professor. And yet, when she died two years later, she was soon forgotten, just as Émilie, too, had soon faded from public memory. Despite the successes of these women (and a small handful of others), most scientific academies remained closed to women, so that intellectually exceptional women continued to be just that: exceptions – a fact Rousseau had exploited, in 1762, to prove his case against granting equal political rights to women. Consequently, although much had changed scientifically and politically by the mid-1790s, mathematically inclined young girls like Mary Somerville had to begin the struggle for acceptance all over again.

In fact, the division between the sexes was actually increasing at this time, partly as a consequence of scientific empiricism itself, particularly the emerging science of biology. As more was discovered about the reproductive differences between the sexes, women became increasingly seen in opposition to men – not simply inferior but completely 'Other' to the masculine human norm. This trend, combined with the rise of male-dominated mathematical physics and rationalist politics, meant reason itself became increasingly synonymous with masculinity, and women were seen not merely as inferior in reasoning ability, but as inherently irrational. This meant intellectual women were increasingly considered by the mainstream to be not so much exceptional as *unnatural.* Mary Somerville felt this trend keenly: she did not believe intelligent women were unnatural, and, as she recalled in her memoir, 'I was annoyed that my [childhood] interest in reading was so much disapproved of [by my relatives and neighbours], and thought it unjust that women should have been given a desire for knowledge if it was wrong to acquire it.'

Somerville was Mary's married name – she was born Mary Fairfax, on 26 December 1780. She grew up in the seaside village of Burntisland, directly opposite Edinburgh across the picturesque estuary known as the Firth of Forth. She was the second of four children (three others had died in infancy), and was particularly close to her older brother, Sam. Her childhood was happy, despite the family's genteel poverty: her father would rise to become Admiral Sir William Fairfax, but his naval pay remained relatively meagre. Nevertheless, Mary enjoyed a reasonable standard of living, because her practical, easygoing mother grew vegetables and kept a cow for milk and butter, and an old orchard provided abundant fruit.

Mrs Fairfax taught her daughter to read the Bible, and insisted she learn the Calvinist catechism, but apart from these studies and her household chores, Mary was allowed 'to grow up a wild creature', as she put it, free to roam among the birds and flowers, or wade in the inlet at the bottom of the garden. Sometimes, when she and Sam sat on the mossy rocks and gazed across the sea to Edinburgh, they would see whales spouting and dolphins playing. But mostly she spent her leisure time alone – Sam was away in Edinburgh during the week, staying with their grandparents so he could attend school, and later, university. Consequently, the wild birds were Mary's special childhood companions, and her love of them stayed with her all her life. When she was in her eighties, she had a pet mountain sparrow, which for many years 'was my constant companion, sitting on my shoulder, pecking at my papers, eating out of my mouth; and I am not ashamed to say I felt its accidental death very much'. This bird had 'both memory and intelligence, and [it had] such confidence in me as to sleep upon my arm while I was writing'. She lamented the steady decline of 'the feathered tribes' in Europe, noting perceptively, 'They will certainly be avenged by the insects.'

Although she had notes to draw from, Mary did not begin working on her memoir until she was nearly eighty years old, with the final draft completed when she was ninety-one – and yet it still sparkles with wonderful anecdotes and historical details that give a sense of immediacy to her writing. In recalling her childhood, she gave vivid descriptions of old Scottish customs that had since died out, and she recalled that 'most of the common people, and many of the better educated', had still believed in the existence of ghosts and witches.

Mary's memoir also gives a brief insight into the political situation in Britain during her youth. She had been a free-spirited child running wild in the Scottish countryside when the French Revolution began in 1789, but the Revolution's shock waves had reached even provincial Scotland, and she later recalled, 'The corruption and tyranny of the court, nobility, and clergy in France were so great that, when the revolution broke out, a large portion of our population thought the French people were perfectly justified in revolting, and warmly espoused their cause.' Such support for the Revolution also reflected the fact that conditions for ordinary people were harsh in Britain as well as France, and Mary wrote, 'At this time, the oppression and cruelty committed in Great Britain were beyond endurance. Men and women were executed for what at present [she was writing in the early 1870s] would only have been held to deserve a few weeks' or months' imprisonment.' (Indeed, in the 1790s, the British were colonising Australia by transporting petty criminals – and political radicals – who were often treated with 'cruelty beyond endurance'.)

Mary added that the British people were sharply divided in their opinions at this tumultuous time, and conservatives accused liberals of fomenting revolution at home:

> Great dissensions were caused by differences of opinion in families . . . My father [and one of my uncles] were as violent Tories as any . . . [but] the unjust and exaggerated abuse of the Liberal party made me a Liberal. From my earliest years, my mind revolted against oppression and

> tyranny . . . [and] my liberal opinions, both in religion and politics, have remained unchanged (or, rather, have advanced) throughout my life . . .

An early example of Mary's commitment to liberal principles was the fact that for some time, she and Sam refused to take sugar in their tea, in protest against the institution of slavery. (With the growth of nineteenth-century Romantic humanitarianism, and the efforts of reformers like William Wilberforce, Thomas Clarkson, and the Quakers, slavery would be made illegal in the British West Indies in 1807, and in the rest of the British Empire in 1833.)

The British government – under the popular William Pitt (the Younger) – was so fearful of liberal support for the French Revolution that it tightened the laws against sedition. It restricted freedom of assembly and freedom of speech (with respect to 'anti-government' opinions), and for a while it went so far as to suspend the Habeas Corpus Act in its bid to stifle radical political debate. The concept of habeas corpus – whereby a person cannot be imprisoned or tried without due legal process – had been invoked in various ways for centuries, but the Habeas Corpus Act of 1679 was the first significant legal codification of this concept. (Incidentally, Locke's support for this act had been politically influential.) As a consequence of Pitt's suspension of the Habeas Corpus Act, there were numerous 'treason trials' in the 1790s. For instance, Mary spoke of the radical reformists John Thelwall, John Horne Tooke and Thomas Hardy: she called these men 'martyrs', who were 'tried for their opinions'. To the government's surprise – and to Mary's joy – they were acquitted, thanks to the skill of their defence lawyer, Thomas Erskine.

Erskine was a supporter of the early French Revolution, and he defended a number of pro-revolutionary radicals who were prosecuted in Britain – including the remarkable, self-taught Thomas Paine. In 1791–92, Paine had published his two-part *The Rights of Man*, a defence of the revolutionary French constitution and its underlying ideals as expressed in the famous *Déclaration des droits de l'Homme et du Citoyen* (*Declaration of the rights of Man and Citizen*). One of the most extraordinary members of Paine's circle was Mary

Wollstonecraft, who, in 1790, had penned her own defence of the French Revolution, *A Vindication of the Rights of Man*. Two years later, when Mary Somerville was eleven years old, Wollstonecraft published her legendary *Vindication of the Rights of Woman*.

Mary Somerville did not mention Mary Wollstonecraft in her memoir, presumably because she was unfamiliar with Wollstonecraft's work, not because she was uninterested in it – after all, she, too, spoke of the rights of women, and of liberal politics. Rather, it seems to be another case of a radical woman's voice being forgotten soon after her death – Wollstonecraft died in childbirth in 1797, when Somerville was a sixteen-year-old tucked away in rural Scotland. (Wollstonecraft's daughter grew up to be Mary Shelley, author of *Frankenstein*.)

Some of Wollstonecraft's most witty and incisive analysis concerned her response to Rousseau's educational blueprint, in which girls' education was designed solely to make them better wives and mothers. Rousseau believed gender roles were 'natural' – an unproven hypothesis that led Wollstonecraft to cite Newton: she said Rousseau's analysis of women was purely to 'indulge his feelings', whereas he should have calmly investigated the cause of things through 'quiet contemplation, like Sir Isaac Newton'. As part of her critique of his shoddy thinking, she turned Rousseau's appeal to nature against him by pointing out that young animals in their natural state require continual exercise – something that girls' upbringing usually precluded. Instead, at school they were 'obliged to pace with steady deportment stupidly backwards and forwards, holding up their heads and turning out their toes, with shoulders braced back, instead of bounding, as Nature directs, in the various attitudes so conducive to health'.

Rousseau had also claimed girls were 'naturally' fond of dolls, clothes and chattering. Wollstonecraft retorted that dolls never excite girls' attention 'unless confinement offers them no alternative'. She added, 'Most of the women, in the circle of my observation, who have acted like rational creatures, or shown any vigour of the intellect, have accidentally been allowed to run wild [as children].'

This was true of Mary Wollstonecraft and her circle, but it was also true of Mary Somerville, a generation later. She spent so much time running wild that she only attended a formal school for one year, when she was ten years old. This was exactly the same time, 1791, that Wollstonecraft was writing her polemic, and when Mary Somerville returned from that dreadful year at school, she 'felt like a wild animal escaped out of a cage', as she put it in her memoir. This was a literal description, because at her prestigious girls' boarding school, she and her fellow students had been forced to study mindless rote learning tasks while being enclosed in a steel contraption that encouraged 'ladylike' posture, forcing the chin up and pulling the shoulders back until the shoulderblades met. As for dolls, she had had little time for them, preferring to play in the garden, getting to know the wild birds and their habits, or running along the seashore, collecting and studying shells, starfish, birds' eggs, seaweed and coastal flowers. This childhood connection with nature was the foundation of her later interest in science.

Like Émilie and Wollstonecraft – and the equally free-spirited Catharine Macaulay and Olympe de Gouges, whose respective *Letters on Education* (1790) and *Declaration of the Rights of Woman* (1791) had inspired Wollstonecraft – Mary Somerville was forthright in her views on women's education. For instance, speaking of fashionable books on the topic – including those written by progressive women like Hannah More – she wrote, 'I detested their books, for they imposed such restraints and duties [on women] that they seemed to have been written to please men.' She added, 'I resented the injustice of the world in denying all those privileges of education to my sex which were so lavishly bestowed on men'.

Of course, it was not only women who agitated for women's rights during the late eighteenth century. For example, Nicolas de Condorcet was a famous mathematician and a disciple of Voltaire, and he was also inspired by his feminist wife, Sophie: in 1790, he petitioned the French Revolutionary National Assembly to grant equal rights to women, declaring: 'Either no individual of the human race has genuine rights, or else all have the same, and he who votes against the right of another, whatever the religion, colour or sex of that other, has henceforth publicly renounced his own rights.' A year later, the constitution was amended to include specific protections for Protestants and Jews, but nothing was added about women, which prompted De Gouges's fiery *Declaration of the Rights of Woman*, a gender-reversed version of the *Declarations of the Rights of Man and Citizen*, in which she famously declared, 'Woman has the right to mount the scaffold [of the guillotine]; she must equally have the right to mount the rostrum [of public and political address].' Tragically, De Gouges and Condorcet were ultimately denied the rostrum and were guillotined for criticising the extreme views of the Jacobins.

All in all, Mary Somerville grew up in a world that excluded women from participating in political and intellectual life, and her story is one of poignant, truly inspiring struggle. But it is also a story of happiness and hope, so much so that even in her own lifetime, her example – and the support she received from scientific men – began to chip away at the negative stereotype about mathematical women.

The first seed of Mary's future mathematical achievements was sown, in a most unexpected way, sometime in the mid-1790s, when she was about fifteen years old. She used to go with her mother to the tea parties given by one or other of the 'widows or maiden ladies' of Burntisland, and she usually found them very boring; on one occasion, however, a younger woman was present, a Miss Ogilvie, who invited Mary to come to her house to see her exquisite needlework. She also showed Mary a monthly magazine that provided her sewing patterns. But Mary was more interested in the fact that the magazine also contained mathematical puzzles and their solutions, because it was there that she saw, for the very first time, 'strange looking lines mixed with letters, chiefly x's and y's'. She found these symbols mesmerising, but Miss Ogilvie could tell her nothing else but that 'they call it algebra'.

Like brain-teasers in today's newspapers, solving the puzzles in eighteenth-century ladies' magazines required a facility with arithmetical concepts such as factors, and knowledge of how to factorise algebraic equations – which shows readers of 'the fair sex' were presumed to have some degree of mathematical education, equivalent to at least Year 9 mathematics today. Recall this was also the level of Émilie's first lessons with Maupertuis. But young Mary Fairfax had no idea of such things. Nevertheless, she was so fascinated by those mysterious x's and y's that she went home and searched her father's library for a book on algebra – but the closest she could find was a work on navigation. She persevered for quite some time, trying to decipher the meaning of the trigonometry in this book: although it did not tell her about algebra, it did arouse her curiosity about the science of astronomy, which, to her amazement, evidently consisted of much more than simple star-gazing.

None of her family or acquaintances had any interest in science. Besides, even if any of her kin *had* had such knowledge, she would not have dared ask for help because they would have laughed at her. And young Mary would have died rather than be laughed at, such was the judgmental society in which she lived. She later recalled, 'I was often very sad and forlorn; not a hand held out to help me.' The only person who did encourage Mary's intellectual development was her uncle, Dr Thomas Somerville of Jedburgh, with whom she sometimes stayed. During one such visit, he invited her to study Latin with him for an hour every morning before breakfast. But he could not tell her about mathematics.

Because of her fear of being laughed at, Mary did not dream of trying to buy or borrow a textbook herself, but when her parents hired a tutor for her younger brother, she saw a way forward at last. She thought the tutor was a 'simple, good-natured man', and she felt confident enough to ask him to buy her, on his next trip to Edinburgh, a copy of the algebra textbook used in schools at that time, and also a copy of Euclid's *Elements* (whose geometrical chapters, or 'books', were widely used in schools for two thousand years, until the early twentieth century). He obliged, and Mary's path was set. Her days were filled with household chores – including, by now, making and mending her own clothes – and with painting lessons and the hours of piano practice expected of young ladies. But at night she stayed up late reading her geometry and algebra books, so much so that it was soon noticed the household stock of candles had run low unusually quickly – after which her candle was confiscated as soon as she was in bed!

With no candles to study by, she spent her nights lying in the dark recalling various proofs from the six books of Euclid she had already read. But when her father came home from sea and heard about her new passion, he affirmed the family position on the matter: 'We must put a stop to this, or we shall have Mary in a straitjacket one of these days.' The widespread presumption that serious intellectual study would damage women's health – because their brains were not capable of high-powered thinking – was a consequence of the belief that such higher study was unnatural for women. But against all these odds, Mary continued to study in secret. Like Émilie and Laura Bassi, she was driven by ambition, which was also considered unnatural in women: 'I was intensely ambitious to excel in something, for I felt in my own breast that women were capable of taking a higher place in creation than that assigned to them . . .'

Mary grew into a fine-looking young woman, and she enjoyed her beauty as well as her brains. But she tried not to become too serious about it: 'I was now a very pretty girl, and much admired, though my mother used to say that her family were like pigs, pretty when young, but grew uglier every day.' In Edinburgh circles, she was called 'the Rose of Jedburgh', because of her delicate features (and because she had been born in her aunt and uncle's house at Jedburgh).

In recounting this period of her life, Mary's *Recollections* become delightfully evocative of Jane Austen's novels: Mary is the provincial relation with no fortune who is often invited to stay with richer or more urban relatives, and she gives simple but fascinating descriptions of the balls, the flirtations, the match-making, the picnics, the evening entertainments where young ladies played the piano or sang – and also the comedy-dramas and the true tragedies. And everywhere there were the gossiping, judgmental friends and relatives – the bane of Austen's heroines' lives, and of Mary's: 'In a small society like that of Edinburgh there was a good deal of scandal and gossip; everyone's character and conduct were freely criticised, and by none more than [one particular] aunt and her friends. She used to sit at a window embroidering, where she could not only see everyone that passed, but with a small telescope could look into the dressing room of a lady of her acquaintance and watch all she did.'

It is no surprise that Mary's descriptions of her youth read like a Jane Austen novel, because Austen was writing at exactly the same time and in exactly the same kind of provincial, middle-class circumstances that Mary was describing: she began writing her famous books in about 1797, when she was twenty-two (and Mary was sixteen). The four major novels published in her short lifetime – she died at forty-one – were published anonymously from 1811–1816, but the posthumously published ones would appear under her name in 1818, and it was around then that Mary first discovered Austen's works: 'I thought them excellent, especially *Pride and Prejudice*. It certainly formed a curious contrast to my old [teenage] favourites, the [gothic] Radcliffe novels and the ghost stories – but I now had more discretion.' In fact, Austen had initially set out to satirise the gothic emphasis on sensibility and melodrama, but she ended up with a unique style in which she made great novels out of everyday life, and her heroines had spirit and common sense rather than Romantic, tempestuous passions. In this regard, they were heroines very much like Mary.

In 1804, twenty-three year old Mary married a distant cousin, Samuel Greig, who was a commissioner in the Russian navy and a London-based Russian consul: Greig had grown up in Russia, his father having gone there as a young British naval officer sent by the government to help Empress Catherine reorganise her navy. Mary said very little of her husband in her memoir; perhaps she married him in the hope of escaping Burntisland and meeting more intellectually open-minded people. But Greig was not a kindred soul: 'I met with no sympathy whatever from him, as he had a very low opinion of the capacity of my sex, and had neither knowledge of nor interest in science of any kind.' Living in London, where she had no friends and was left alone all day while Greig was working, she began to teach herself French, and to study higher mathematics; presumably she had been able to buy herself some more advanced textbooks to guide her study.

After three years of marriage, however, Greig died at the age of twenty-nine. Mary did not say how he died – and I can find no other record of it – but she did say that after his death she was 'much out of health'. She also had two young sons to care for, one of whom she was still nursing; the older boy was named Woronzow, in honour of the Russian Ambassador to Britain.

When Mary returned to her parents' house after her three years with Greig, she had a small inheritance from her husband and the respectability of widowhood. She also had a new sense of independence and confidence – so much so that she began to read Newton's *Principia*, although she soon put it aside because it was so difficult. She would return to it later, but at that early stage it is a wonder she tried to read it at all: she had a toddler, and a baby she was still breast-feeding, and she was expected to help run her parents' household. She must have had a truly extraordinary determination to enter into the mysterious world of mathematics, and it is incredible to think how long and hard she had already persisted with so little help. But everything was about to change and, once again, inspiration came in the form of a magazine.

When the requisite period of mourning was over and Mary returned to society, she met with the liberal intellectuals associated with the literary *Edinburgh Review*, including Sydney Smith and Henry Brougham. She also met the venerable John Playfair, Professor of Mathematics and Natural Philosophy at Edinburgh University. Then in his sixties, Playfair still enjoyed female company – and the ladies liked him. Mary appreciated the fact that he encouraged her to persevere with her solitary study of higher mathematics. A decade later, Playfair would praise Émilie's translation and commentary on the *Principia*, in a historical overview of science he was asked to contribute to the *Encyclopaedia Britannica*. But Mary had moved back to London by then, and she does not seem to have known of Émilie's existence.

Meanwhile, through her Edinburgh circle, Mary managed to obtain a mathematical journal published by a protégé of Playfair: William Wallace, who would later become Professor of Mathematics at Edinburgh University. His journal, the *Mathematical Repository*, contained mathematical puzzles that readers were invited to solve;

recalling her excitement at seeing the puzzles in Miss Ogilvie's copy of a women's magazine, Mary tried her hand at Wallace's more sophisticated puzzles. Wallace was a self-taught mathematician himself: he had started out as an apprenticed bookbinder, and had met his mentor, Playfair, in the bookshop where he worked. When he received the solutions Mary submitted, Wallace kindly sent her his own solutions as a guide for her study; this led to a correspondence in which, for the first time in her life, she received some serious mathematical instruction.

The *Mathematical Repository* offered prizes for correct solutions to the more difficult puzzles, and in 1811, when she was thirty years old, Mary was thrilled to win a silver medal. At last, she had 'arrived' publicly as a legitimate aspirant to the title of 'mathematician'. Inspired by her success, she continued her self-directed study of higher mathematics and astronomy, but because of her lack of formal education, she could not escape the nagging feeling that perhaps she did not properly understand what she was reading. At Wallace's suggestion, she engaged a tutor to read with her, and the book she chose to study was Laplace's *Mécanique Céleste* (*Celestial Mechanics*), which was an extraordinarily ambitious choice. Furthermore, she had to read it in French. Amazingly, she discovered that she seemed to understand Laplace as well as her tutor did, which 'gave me confidence in myself and consequently courage to persevere'.

Laplace's aim in his massive, five-volume *Celestial Mechanics* was to summarise the theory of gravity in modern mathematical language, and to collect in one place all the related mathematical results that had arisen since Newton published his 'admirable' work (as Laplace called it). In the fifty years between the completion of Émilie's commentary and the publication of the first volumes of Laplace's masterpiece, an enormous amount of new mathematics had been created from Newton's fertile material, which meant Mary had attained an incredible degree of mathematical sophistication to be able to understand *Celestial Mechanics.* Most of this new mathematics had been created by Émilie's colleagues – Bernouilli, Euler, Clairaut, D'Alembert and Maupertuis – and by Mary's future French colleagues, Lagrange and Laplace; relatively little new mathematics had come from Britain. I say 'future colleagues' because Mary's newly confident reading of *Celestial Mechanics* would ultimately lead to her meeting Laplace himself. But that would come later, because Mary was about to fall in love. In 1812, the year after she won her silver medal, she married the man who would prove to be her soul mate and constant support: forty-one year old William Somerville.

William was her cousin, the son of her favourite aunt and uncle, the Somervilles of Jedburgh. Trained as a medical doctor, he had spent many years abroad. In South Africa, for example, he had held a diplomatic post in which he was responsible for trying to make a treaty with the indigenous tribes, and in so doing he had become the first white man to reach the Orange River. He also had an 'illegitimate' son, whom he brought back to England, and who would become quite close to Woronzow, Mary's son from her first marriage. William, too, had been married before (some time after his earlier liaison), but his wife and their infant son had died.

When Mary met up with William in late 1811, she had not seen him since childhood; she was immediately attracted to his 'liberal principles', and to the facts that 'he had lived in the world, was extremely handsome, had gentlemanly manners, spoke good English, and was emancipated from Scotch prejudices'. (By contrast, one of her other suitors at this time had had the 'impertinence', as she later put it, to send her a book of sermons 'with the page ostentatiously turned down at a sermon on the Duties of a Wife, [which was written] in the most illiberal and narrow-minded language'.) But it was on her marriage to William that Mary realised just how much her family had disapproved of her studious ways – excepting, of course, William's father, her kindly uncle Thomas, and belatedly her own parents. For instance, she received a letter from William's sister, saying she hoped Mary would give up her 'foolish manner of life and studies, and make a respectable and useful wife for her brother'. Although Mary was 'extremely indignant' at this judgmental comment, she and William would never look back.

## *Chapter 14*

# THE LONG ROAD TO FAME

Mary and William spent their honeymoon in the Lake District, childhood home of the Romantic poet Wordsworth. Wordsworth famously immortalised the wondrous time in France when there was dancing in the streets because the French Revolution promised to be the dawn of a new era of freedom: 'Bliss was it in that dawn to be alive, but to be young was very heaven!' During the Terror, he retreated to the peace of the English countryside and sought inspiration from nature instead of politics. British Romantics tended to believe in an innate opposition between poetic sensibility and scientific rationalism: they felt that, in divesting nature of its mystery, science had destroyed something precious in the human spirit and imagination – or, in Keats's memorable words, it could 'unweave a rainbow'. This division between the humanities and sciences still lingers in Britain and other English-speaking countries, but it was never so strong on the Continent. For instance, Johann Wolfgang von Goethe, the German writer, biologist and sometime Romantic, believed that in expressing the unity of nature, science and art could themselves be unified. Goethe was thinking more of biological science, however: he objected to Newtonian physics because he believed it to be too mechanical and lifeless.

Mary disagreed with this kind of pastoral, Romantic, anti-Newtonian philosophy. She felt that studying the heavens mathematically was 'sublime': it stretched rather than constrained the imagination, and it was awe-inspiring to contemplate how mathematics had led to so much more being discovered about the universe than anyone would have believed possible. In the introduction to her first book, she would write with wonder of the incredible economy and power of the Newtonian laws of physics: 'The infinite varieties of motion, in the heavens and on earth, obey a few laws so universal in their application that they regulate the curve traced by a [leaf], which seems to be the sport of winds, with as much certainty as the orbits of the planets.'

She was also alive to the numinous side of astronomy: 'The magnitude and splendor of [celestial] objects, the inconceivable rapidity with which they move, and the enormous distances between them, impress the mind with some notion of the energy that maintains them in their motions with a durability to which we can see no limit.'

Equally awesome to her was the fact that humans have the faculties to discern and appreciate these creations of 'the great First Cause', although she felt that contemplating the cosmos also 'inculcates humility, by showing that . . . however profoundly we may penetrate the depths of space, [the universe is so vast that we] dwindle into insignificance . . .' (Like Newton, and also Émilie and Voltaire, Mary publicly adopted a deist, 'First Cause' position in her writings in order to focus on science as a secular discipline. Personally, she had long ago left behind what she called the 'gloomy doctrines of Calvinism', but she remained a Christian, and continued to find comfort in the Bible. However, she and her husband kept their religious views to themselves, and did not enter into religious controversy with any of their friends, because 'we had too high a regard for liberty of conscience to interfere with anyone's opinions'.)

It was these profound aspects of mathematical astronomy that inspired Mary to pursue her studies through such difficult and drawn-out circumstances – and true to her passion for the sublime science of gravity, when she first returned to study after her honeymoon, she took up both *Celestial Mechanics* and the *Principia*. She had put aside the latter years before, because 'I found it extremely difficult, and certainly did not understand it till I returned to it some time after, when I studied that wonderful work with great assiduity'. She was helped in her study by the commentary in Jacquier and Le Seur's Latin edition of the *Principia*: recall Jacquier had inspired Émilie to embark on her French translation but, as indicated earlier, it seems Mary did not know of Émilie's later edition and commentary.

At this time, Mary also bought books by Lagrange, Clairaut and Euler, and she later recalled, 'I was thirty-three years of age when I bought this excellent little library. I could hardly believe that I possessed such a treasure when I looked back on the day that I first [heard] the mysterious word "Algebra", and the long course of years in which I had persevered almost without hope. It taught me never to despair.'

It was a lesson she would need to remember, because although her life with William was a fortunate one in the main, nevertheless she did have some cause for despair: in particular, there was a disastrous financial loss (the result of a bad debt by a relative she and William had trusted, and for whom William had gone guarantor), and worse, there were the early deaths of three of her six children: in the same year that she bought her library of French books, her second son from her first marriage died, at the age of nine, and a year later, her first son with William died when still a baby. At that time, she and William also had a toddler daughter, Margaret, and their second daughter, Martha, was born the following year, 1815; their third daughter, Mary, was born in 1817. But in 1823, their beloved Margaret died at the age of ten, and Mary fell into despair at the thought that perhaps rigorous study *was* unnatural for girls: decades later, she wrote, 'I feared I might have strained her young mind too much.' This was a reference to the fact that, while her sons went to school (and Woronzow, her only surviving son, would go on to Cambridge), Mary herself taught her daughters algebra, Latin and Greek, for several hours a day.

William Somerville, too, helped check the children's homework: he was very fastidious in writing English, and was a useful if severe critic of the children's work and, later, of his wife's manuscripts. Temperamentally, he was outgoing and sociable, and although he was keen to provide well for his family through his medical work, he had no personal ambition as a scholar or writer. Instead, he was content to help Mary with her research, and with the laborious task of copying her manuscripts for distribution. Their daughter Martha recalled, 'No trouble seemed too great which he bestowed upon her; it was a labour of love.' This was not just daughterly affection: Mary herself recorded that, whereas her early studies had been constantly beset with 'every difficulty', now she received 'every encouragement', thanks to her husband's 'kindness and liberal opinions'.

The Somervilles soon became part of the inner circle of scientific life, first in Edinburgh and then in London, when William took up a medical appointment there. One of their earliest and most memorable English connections was made when Mary's first mentor, William Wallace, introduced them to the venerable F. William Herschel; recall Herschel had carried out the experiment anticipated by Émilie in her essay on fire, proving that different parts of the solar spectrum do, indeed, express different amounts of heat.

The German-born Herschel and his sister Caroline worked together, as is well known – he discovered the planet Uranus, and she discovered a number of new comets and nebulae. In 1800, in the course of his telescopic observations, Herschel had noticed that different coloured filters produced light that gave different sensations of heat, just as Émilie had suspected. To test whether this was a property of the filter or the light itself, he used a prism to create a spectrum from sunlight, and placed a thermometer at different parts of the spectrum. Émilie had not tried this simple experiment, presumably because she assumed the temperature differences for the different colours would be so small it would be difficult to 'gather enough rays' of each colour, as she put it; or perhaps it was simply because she could not borrow Voltaire's thermometers without alerting him to the fact that she was writing her own essay. Surprisingly, Herschel found it took only a few minutes' exposure to the spectrum for temperature differences to become noticeable. He found, as Émilie had anticipated, that red light is the warmest, while violet is the coolest.

Herschel also noticed the temperature was even higher just outside the red end of the spectrum – and thus he discovered the existence of infrared rays.

It was Herschel's astronomical discoveries that had captured the public imagination, however, and the Somervilles were delighted to visit the huge reflecting telescope he had built. For his part, the seventy-four year old Herschel had been looking forward to meeting Mary, because he 'highly esteemed' the fact that she had become a good mathematician, according to Wallace, without having graduated from Cambridge. Naturally, Mary was thrilled to meet him, too, although she was disappointed Caroline Herschel was away at that time. But it was Herschel's son, John – ten years younger than Mary, and still 'quite a youth' – who would become one of her most important mentors. (Caroline would return to Germany when her brother died in 1822.)

With William Somerville's broad interest in science and Mary's specialist knowledge of mathematics, and with her charm and his gregariousness, the Somervilles were very popular in scientific society, so much so that Mary later recalled she had met with 'nothing but kindness' from scientific men. These men included not only Wallace and the Herschels, but also William Wollaston, Henry Kater, Thomas Young, Michael Faraday, Charles Babbage and Alexander Marcet in Britain, and François Arago, Jean-Baptiste Biot, Laplace and Siméon Denis Poisson in France. These men's wives were also part of the social circle, and some of them, notably Mrs Kater and Madame Biot, helped their husbands with their scientific work, while Jane Marcet had made her own reputation, with her husband's encouragement, as a scientific populariser.

Mary developed warm friendships with most of these women and their husbands, and she delighted in the camaraderie and excitement of social gatherings with her new friends. Later on, she and William often entertained at their house, but in their early London days they frequently spent evenings at the Katers', along with Wollaston, Young and others. 'Sometimes we had music, for Captain and Mrs Kater sang very prettily. All kinds of scientific subjects were discussed, experiments tried, and astronomical observations made in a little garden in front of the house.'

One evening, the Somervilles stayed late at the Katers', looking through a telescope until two in the morning. On the way home, they saw a light in Young's window, so they decided to call in and see him. He was in his dressing gown, but he said, 'Come in, I have something curious to show you.' Young was a polymath: only seven years older than Mary, he was originally trained as a medical doctor, but in the first decade of the nineteenth century, he had turned to physics. He was also well known for his work in a completely different field, Egyptology, and it was this work he wanted to discuss with the Somervilles that morning: he had just deciphered a horoscope on an Egyptian papyrus, and had used his knowledge of astronomy to date the papyrus to the early days of Alexandria. He did this by comparing the configuration of stars represented on the papyrus with modern astronomical data.

Young had also helped unlock the code of Egyptian hieroglyphics through his study of the famous Rosetta stone – a black rock with two-thousand-year-old inscriptions in ancient Greek, 'spoken' Egyptian, and Egyptian hieoroglyphics. The Rosetta stone had been stolen by one of Napoleon's soldiers in 1799, during the French occupation of Egypt. But Young is most famous for the fact that, in 1801, he had discovered new evidence that light appears to behave as a wave, not a particle as Newton had suggested. Mary was an early supporter of Young's wave theory of light, and she would describe it, and how it differed from Huygens's wave theory, in her second book.

Meanwhile, she revelled in these spontaneous associations with her new colleagues. Another such example was a surprise visit from Wollaston, who had become her close friend and mentor. Like Young, he had started out as a medical doctor but had turned to physics and chemistry, fields in which he made numerous discoveries – including the one he demonstrated at the Somervilles' house that day. It was a sunny morning, and Mary recalled, 'Closing the window-shutters so as to leave only a narrow line of light, he put a small glass prism into my hand, telling me how to hold it.' And then she saw it: seven dark lines punctuating the solar spectrum. Newton had shown that sunlight passed through a prism produces a continuous spectrum, like a piece of rainbow, but Wollaston was the first to notice that sometimes a spectrum is marked by an *absence* of light – that is, by tiny black lines. Mary explained it was only possible to see these lines when the prism was as pure as possible, and the beam of sunlight very fine, and that 'the best method is to receive the spectrum on the object glass of a telescope, so as to magnify [the lines] sufficiently to render them visible'.

What Wollaston had discovered was an 'absorption spectrum': the black 'spectral lines' indicated wavelengths of light that must have been absorbed before the light beam reached the eye. At that time, it was not known just *what* was absorbing this missing light, but the answer lay in the later discovery (by Gustav Kirchhoff) of analogous 'emission spectra'. When a gaseous element like sodium or neon is heated or electrically stimulated, it produces light – as in sodium (yellow) street lights and neon advertising signs; at low pressures, the spectrum produced by this light is characterised by tiny lines of colour at specific wavelengths (or positions on the spectrum). Each element produces a unique arrangement of these bright 'emission' lines, which correspond precisely to the dark lines of the absorption spectrum of the same element. This suggested it was 'cooler' gaseous elements in the outer layers of the sun that were absorbing certain frequencies of sunlight and causing Wollaston's dark spectral lines.

Wollaston's and Kirchhoff's discoveries were the beginning of the new field of spectroscopy: knowing the signature emission and absorption spectra of various elements means the composition of gases can be deduced simply from their light spectra; for instance, Mary pointed out that the nature of the gases in the sun and other stars, and in cosmic nebulae – as well as the existence of new elements on earth, including helium – were discovered through spectral analysis. (In the twentieth century, emission and absorption spectra played a key role in the development of quantum theory – which was pioneered by Max Planck, Einstein and Niels Bohr – and today, physicists explain these spectra as follows: when a gas is heated its atoms gain kinetic energy, some of which can be absorbed by the atoms' electrons, raising them to specific permitted higher-energy levels. In a short time, such electrons can 'jump' back to their lower-energy levels, and the extra energy is emitted as photons, with specific wavelengths of coloured light corresponding to the difference of those energy levels.)

The importance of Wollaston's discovery, and the wonder with which Mary responded to his demonstration, meant she misremembered the dates: her *Recollections* suggest he had just discovered spectral lines the day he came to her house – which would be some time after she and William moved to London in 1816 – but in fact he had made his discovery in 1802. Fraunhofer made the same discovery independently in 1814, so perhaps recent news of Fraunhofer's extensive catalogue of spectral lines had prompted Wollaston to show Mary this fascinating phenomenon. Mis-remembered or not, the incident shows the importance of her involvement in scientific society: she had studied mathematics alone, but now, by mixing with such an exciting group of scientists, she also gained an education in experimental physics.

Mary's circle included leading physicists on both sides of the English Channel. The Somervilles first met Biot and Arago in London in 1816, where the Frenchmen were engaged in measuring distances along a meridian of longitude, just as Maupertuis and Clairaut had done. As Mary would note in her first book, if the earth were a perfect sphere, its circumference could be calculated by physically measuring the actual length of a degree along a single meridian, and then multiplying this distance by 360, the number of degrees in a circle. This is essentially what Eratosthenes had done two thousand years earlier, when he made the first estimate of the circumference and radius of the earth. (A thousand years later, tenth-century mathematician al-Birumi improved Eratosthenes's calculation by making more accurate land measurements using triangulation, but he still assumed the earth was spherical.) But, as Mary pointed out, the earth is oblate – a fact taken for granted since Maupertuis's Arctic voyage – so Eratosthenes's method was not sufficient: it was necessary to measure different meridians at different latitudes in order to make more precise measurements of the size of the earth, which is why Biot and Arago were in London.

Like their British colleagues, the two Frenchmen were involved in numerous different scientific fields, not only longitude and astronomy, but also the study of gases, and of electricity and magnetism. Arago also worked in optics, and, together with another French colleague, Augustin Jean Fresnel, he worked with Young on the wave theory of light.

Mary's only claim to fame at that stage was her curiosity value as a charming woman who could converse on deep mathematical topics, including Laplace's famous *Celestial Mechanics*. But Biot was keen for her to come to Paris and to meet his intelligent, scholarly wife, and the following year, 1817, she and William made an extended visit to the Continent, beginning in Paris. Soon after their arrival, Madame Biot showed the Somervilles the Louvre and the library of the Institut de France, where, Mary recalled, they admired the statue of Voltaire.

One of the first things that struck Mary on this, her first trip abroad, was the apparent lack of intellectual interests among most of the women she met in Paris. Twenty-five years earlier, Mary Wollstonecraft, too, had been shocked by the oppression of French women, whose intellectual curiosity she believed was stifled by their socially enforced 'natural' role as dutiful wives and mothers, as defined by Rousseau. However, Wollstonecraft was a rare exception in Britain, too, and her French contemporary, the redoubtable salonnière and writer, Marie-Louise Germaine de Staël, had felt it was *English* women who were denied opportunity for public intellectual and political discussion. It seems these two exceptional women did not meet: De Staël was in England in 1792–93 whereas Wollstonecraft was then in Paris, observing the Revolution first-hand. As for Mary Somerville, in 1817 she found that Madame Biot was the only woman she met who knew anything of science, and that in general, 'dress is a great object among the French ladies and forms a frequent subject of conversation'. This suggests that Mary did not know of the existence of an extraordinary Parisian woman working in mathematics at that very time: the remarkable Sophie Germain.

Sophie Germain is probably the first truly original female mathematician in history. Just four years older than Mary, she had had a similar struggle to educate herself. Recall it was a time when women's brains were seen as incapable of higher study, and Sophie's parents, like Mary's, took away her candles and the fuel for her fire, fearing her long hours of study would ruin her health. Like Mary, she persevered, getting up early in the freezing cold and wrapping herself in blankets. After repeatedly discovering her like this, with her fingers and her ink frozen but her mind immersed in her study, her parents eventually relented and allowed her to study in more comfort.

Sophie's passion for mathematics had germinated by chance, just like Mary's. In a book in her father's library, she read the story of Archimedes and how, during the siege of Syracuse, he was so engrossed in his mathematics that he did not notice the Roman soldier who approached and demanded his surrender. Misinterpreting Archimedes's absorption for contempt, the soldier killed him, and Archimedes died as he had lived: doing mathematics. In the fearful days of the French Revolution, the teenaged Sophie had been so inspired by Archimedes's passion she decided that she, too, would become a mathematician. This story comes from the preface to an 1896 edition of Sophie's letters and papers, and whether it is literally true or not, she did not marry or have children, but devoted her life to mathematics. She was no dry pedant, however, because she also wrote on the state of sciences and letters, and on the 'common sentiments' they inspire.

Sophie's mathematical journey was inextricably bound up with the fortunes of science during the Revolution. Under the anti-elitist, anti-science ('back to Nature') ideals of Robespierre, the Jacobin government had abolished the Paris Academy of Sciences. But two months after the fall of Robespierre and his faction, who were guillotined in July 1794, the chemist Antoine Fourcroy gave an impassioned speech on behalf of reason and the need for scientific education. He proposed the establishment of a new scientific educational institution, which opened in 1795 as the École Polytechnique. The committee that established the École Polytechnique also included the apolitical Laplace, and the Jacobin mathematician Gaspard Monge, which – given Fourcroy's anti-Jacobin speech – illustrates how mathematicians of different political persuasions still managed to work together during the tumultuous aftermath of the Revolution. It also illustrates that not all Jacobins were anti-scientific.

It is worth digressing here to discuss the international benefits of another example of the way mathematical men of varying political persuasions worked together through the Revolution – namely, the Committee on Weights and Measures, which was established in 1790, and which included Laplace and Condorcet. The Committee suffered some political casualties (including Condorcet), but finally, in 1799, it presented the metric system now widely used throughout the world. Most of us take for granted everyday units of measurement like kilograms and kilometres, but, as Maupertuis's Arctic expedition illustrates, it requires a great deal of effort and scientific expertise to make accurate physical measurements and to find useful and accurate measuring standards: recall Maupertuis's team had measured length in terms of a standard pole they cut from the local fir trees, while in the *Principia*, Newton generally used relative measurements.

After much deliberation, the metre was defined to be one ten-millionth of the distance along a meridian from the equator to the North Pole. This distance was physically represented by a 'metre bar' made of a platinum-based alloy and kept at a precise temperature in the Bureau of Weights and Measures near Paris; accurate copies were sent to international laboratories so the metre could be universally applied. (Even non-metric systems use the standard metre: for example, an inch is defined today as exactly 2.54 centimetres.) The kilogram was defined to be the mass of a standard platinum-iridium cylinder kept at the Bureau. (Time, of course, has long been defined in terms of the earth's daily 24-hour rotation, but over the past thirty or forty years, both the second and the metre have been more accurately defined in terms of more precise measurements than those possible in terms of the earth's rotation and circumference, and I have given details in the Appendix.)

Accurate units of measurement make it much easier to replicate and generalise experimental results, especially when making delicate experiments. Nevertheless, Antoine Lavoisier, another member of the Committee on Weights and Measures, is generally considered to be the founder of modern chemistry because, in the years leading up to the Revolution, he had begun the process of turning chemistry into an exact science by using extremely sensitive scales in his experiments, and by keeping careful measures of the masses of substances before and after chemical reactions. In this way he showed that total mass is conserved during chemical reactions, a fact that helped establish the law of conservation of mass.

Lavoisier and his wife (who was also his scientific assistant) were moderate revolutionaries, but, during the last frenzied weeks of the Terror, Lavoisier's financial and organisational involvement in a company that had collected taxes for the king led him to the guillotine. In trying to intercede on her husband's behalf, Marie Lavoisier spent two months in prison herself, and the judge who condemned him is reported to have said, 'We have no need of scientists.' A year later, however, Fourcroy began teaching Lavoisier's chemistry at the new École Polytechnique. Other professors included one of the greatest mathematicians of the age, Joseph-Louis Lagrange. (Lagrange's response to the news of Lavoisier's execution was succinct and powerful: 'It took them only a moment to cause this head to fall, and a hundred years perhaps will not suffice to produce its like.')

Sophie Germain was eighteen when the École Polytechnique opened, but only men were allowed to study there. When she was about twenty-one years old, however, she managed to obtain copies of Lagrange's calculus lecture notes, through a friend, Antoine-Auguste Le Blanc. After studying Lagrange's course with great excitement, she did something extremely audacious: at the end of the semester, when students who were formally enrolled in classes were invited to submit comments to their lecturers, she wrote to Lagrange, pretending to be Monsieur Le Blanc, who had long since dropped out of the course. Lagrange was so impressed with her comments and questions – which seemed so superior to the work he had seen earlier from the real Le Blanc – that he asked for a formal meeting. This unexpected response threw Sophie into a panic, but she need not have worried, because when he discovered her identity, Lagrange proved to be a helpful mentor. Thus inspired, Sophie continued her private study of advanced calculus, and she also read with enthusiasm a new book on number theory, the *Disquisitiones Arithmeticae*, by German mathematician Carl Friedrich Gauss, one of the best mathematicians in history.

By November 1804, Sophie had mastered enough of Gauss's book to write to him at the University of Göttingen, although she did so under the name of M. Le Blanc, and she described herself as an 'amateur enthusiast'. She told Gauss she was interested in applying his methods to proving Fermat's notorious 'last theorem', which tantalised mathematicians for 350 years, until Andrew Wiles proved it in 1994. The theorem says there are no positive whole numbers $x$, $y$ and $z$ for which equations of the form $x^n + y^n = z^n$ are true, where $n$ is *any* whole number that is greater than 2. The equation *is* true when $n = 2$: for instance, $3^2 + 4^2 = 5^2$. And when $n = 1$, the equation simply reads $x + y = z$, for which there are an infinite number of possibilities: for instance, $1 + 2 = 3$.

Gauss was very pleased to find someone who actually understood his new book, and he responded warmly to 'M. Le Blanc'; the two of them exchanged several letters over the next couple of years.

These were the early years of the Napoleonic empire. In the political vacuum created by the fall of Robespierre a decade earlier, Napoleon Bonaparte was one of the men who rose to political prominence. His strong leadership and innovative legal reforms, after the dreadful years of Terror and faction fighting, led to his being voted emperor of France in 1804. But his increasing role as a virtual dictator was a long way from the early Republican ideal of democratic government, although perhaps not so far from Robespierre's autocratic version of Rousseau's political theory.

In 1806, Napoleon's army defeated the Prussians at Jena, not far from where Gauss lived. Sophie was afraid for his safety, and – perhaps mindful of Archimedes's fate during such a time of war – she prevailed upon a family friend, who was a military general, to protect her mentor. The general sent an officer to check on Gauss, who was most perplexed by such a visitation from an enemy soldier, and even more perplexed when the officer said he had been sent on behalf of Mademoiselle Sophie Germain.

Sophie was immensely relieved to hear news from the general that Gauss was safe. At the same time, Mary, too, had been worried about Napoleon's forces. Although the British had won a decisive victory against the French at Trafalgar in 1805, Mary recalled in her memoir that the British people were terrified by the ongoing threat of invasion: Napoleon was not defeated until 1815, only two years before Mary's visit to Paris. During that frightening time when Napoleon was expanding his empire, the British also had to cope with problems at home, as Mary described:

> There had been bad harvests, and there was a great scarcity of bread; the people were much distressed, and the manufacturing towns in England were almost in a state of revolution; but the fear of invasion kept them quiet . . . [But] although I should have been glad if the people had resisted oppression at home, when we were threatened with invasion I would have died to prevent a Frenchman from landing on our coast. No one can imagine the intense excitement which pervaded all ranks at that time. Everyone was armed, and notwithstanding the alarm, we could not but laugh at the awkward and often ridiculous figures of our old acquaintances, when at drill in uniform.

When Sophie revealed her true identity to Gauss, she apologised for having deceived him, saying she had done so because she 'feared the ridicule attached to the *femme savante*' (which shows that nothing had changed for intellectual women since Émilie's time, more than half a century earlier). Gauss was amazed: he said Sophie was proof of something he previously would have found difficult to believe possible – a woman with a taste and aptitude for 'the mysteries of numbers'.

In the end, Sophie did not publish any of her work on Fermat's theorem, although Laplace's colleague Adrien-Marie Legendre credited her with a result she had communicated to him, and which he used in his proof of the $n = 5$ sub-case. In 1811 and 1813, however, she had anonymously entered essays in the Academy of Sciences' competition on a completely different topic: the mathematical analysis of vibrating surfaces and the theory of acoustics. (Because of the rudimentary state of knowledge of these cutting-edge topics, sometimes the Academy repeated its essay topics over several years.) The mathematical analysis of one-dimensional vibrating strings, such as a plucked violin string, had been pioneered by Émilie's colleague D'Alembert, but Sophie's essay was on two-dimensional vibrating *surfaces* such as a drum (or an eardrum), which required more complicated calculus techniques than those used for the vibrating string.

Unfortunately, Sophie's self-education showed, and her work in both cases was severely criticised by the Academicians, although in the second case, she was awarded an honourable mention. In 1816, however, she won the prize with her third paper on the subject.

It was an extraordinary achievement, and yet, for all that Sophie was publicly marvelled at, hailed as a genius for her prize-winning essay and invited to attend meetings at the Academy of Sciences – the first woman to achieve such an invitation – nevertheless, she remained an outsider, a freak like all high-achieving women: her winning essay was not even published at the time. Predictably, she was often praised at other women's expense: in an article in the *Journal des Savants* in 1817, the same year Mary was in Paris, Biot wrote that Mademoiselle Germain was 'probably the person of her sex who has penetrated most profoundly into mathematics, without excepting [even] Madame du Châtelet, because here there was no Clairaut'. (Clearly the legend that Émilie's work was derivative of the 'great men' in her life had passed into the next century.) Not that Biot meant to offend women by denigrating Émilie's achievements – in fact, Mary found him very supportive. Besides, it is true that Sophie's work was entirely original, whereas Émilie (and Mary) focused on elucidating the work of others. However, just a few months after he published his article in praise of

Sophie, Biot organised a dinner party to introduce the Somervilles to Laplace and Poisson, but, judging from Mary's memoir and other correspondence, Sophie was not a guest at this or any of the dinner parties held for the Somervilles. The two women do not seem to have met, or even to have known of each other's existence.

Perhaps Mary's new colleagues simply did not realise that, as outsiders, she and Sophie would have benefited from each other's company. (Such a meeting would have shown Mary that women were capable of original mathematical creativity, something she had come to doubt.) Or perhaps Sophie did not feel comfortable in the company of the Academicians, some of whom were not very supportive of her work: Poisson, in particular, was working in the same field (on the mathematics of vibrating surfaces), and he had been one of the essay judges who criticised her work most harshly.

In 1831, the very same year that Mary would publish her first book, Sophie Germain died of breast cancer at the age of fifty-five. To make her premature and painful death even more tragic, she died just before she was about to receive an honorary degree from Göttingen, organised by Gauss, whom she had been about to meet in person, at last. It was a sad ending for a woman who had given her life to mathematics. Half a century later, her native city would belatedly pay tribute: in the 1880s, a Parisian street was named after her, and a girls' school: both the Rue Sophie Germain and the Lycée Sophie Germain still exist as a memorial to her.

During the Somervilles' Parisian visit in 1817, Laplace invited them to spend the day at his country house, where they dined with many of their new scientific friends. Mary was seated next to Laplace, whom she found very kind and attentive, and she recalled, 'In such an assemblage of philosophers I expected a very grave and learned conversation. But not at all! Everyone talked in a gay, animated and loud key, especially M. Poisson, who had all the vivacity of a Frenchman.' She sometimes found the conversation in rapid French difficult to follow, but on scientific topics she was more comfortable, since most of the mathematical physics she studied was written in French. William, on the other hand, had spent many years in Montreal, and he spoke fluent French.

A few days later, Mary and William called on Madame de Laplace, and were amazed when the marquise received them from her bed, where she lay 'elegantly dressed. I think the curtains were of muslin with some gold ornaments, and the coverlet was of rich silk and gold. It was the first time that I had ever seen a lady receive in that manner. Madame de Laplace was lively and agreeable; I liked her very much.'

The Laplaces had only recently been elevated to their titles of marquis and marquise, thanks to the new king, Louis XVIII. Napoleon had lost power when he lost his European Wars, his final defeat being at Waterloo in 1815. The monarchy was fully restored by 1816, under Louis XVI's brother, Louis XVIII. Louis XVI had been executed during the Terror, as had his wife, Marie-Antoinette, while their son, young Louis XVII, had died in prison. The new king attempted to run a moderately liberal government, with only mixed success; after a series of uprisings, the French monarchy would finally be dissolved in 1848, and after the rise of another empire under Napleon's nephew, Napoleon III, the modern French republic would begin life in 1870, nearly a century after the Revolution.

Mary continued to exchange occasional letters with Laplace for the next few years, until his death in 1827. Three years earlier, when he was seventy-five years old, he had finished writing the fifth volume of *Celestial Mechanics*, and he wrote to Mary that while he was completing this work, he had re-read the 'incomparable' *Principia*. He said that although he appreciated the 'elegance' of Newton's geometrical proofs, he recognised the need for the modern algebraic form of calculus. After all, it was this 'Leibnizian' calculus that Clairaut had used when working out the perturbations in the moon's orbit – and which Lagrange, and Laplace himself, had used to analyse similar distortions in the planetary orbits.

In fact, the French had made such great progress in solving difficult gravitational problems that it was clear British mathematics had fallen far behind. Some scholars say it was Newton's long shadow that stifled innovation among British mathematicians, and also that his version of calculus was not as conducive to progress as Leibniz's. Not all historians agree with this argument, but it was the view Mary and her colleagues held. Some of them – notably George Peacock, Charles Babbage and John Herschel – publicly called for a reform of British scientific education and practice, including the adoption of the Continental rather than the Newtonian form of calculus. (Babbage is famous for designing the first programmable computer, which he called an 'analytical engine'; at one point, he was assisted by the remarkable Ada Lovelace, estranged daughter of the poet Byron. She married a friend of Mary's son, Woronzow, and she became Mary's mathematical protégée.)

In this contentious climate of criticism of British mathematical science – in early 1827, the year of Laplace's death and the centenary of Newton's – William Somerville received a letter that would change Mary's life. It was a request from Lord Brougham (who had been a co-founder of the liberal *Edinburgh Review*), asking Somerville to prevail upon his wife to write a popular account of Laplace's *Celestial Mechanics*. Brougham wanted something that might be of interest to mechanics and other technical workers, but something not too scholarly. 'In England there are now not twenty people who know this great work, except by name: and not a hundred who know it even by name,' wrote Brougham to Somerville, and he added there was no-one besides Mary who was qualified to write such a work. Mary was stunned: 'This letter surprised me beyond expression. I thought Lord Brougham must have been mistaken with regard to my acquirements, and naturally concluded that my self-acquired knowledge was so far inferior to that of the men who had been educated in our universities that it would be the height of presumption to attempt to write on such a subject, or indeed on any other.'

However, at that time, anyone who had learned the French methods in Britain had learned them by reading French books, as Mary had done, not from British universities, and Brougham later paid a personal visit to try to persuade her to take up his challenge. She eventually agreed to his proposal on two conditions: firstly, any attempt to explain the applications of the Continental form of calculus, such as Brougham had requested, would place the book beyond the popular audience, so she would be writing a more serious work than originally proposed. Secondly, she would undertake the work in secret, and if she failed to produce anything worthwhile, the manuscript would be burned. And so began Mary's labour of love; like Émilie's 'Newton', it was fitted into the cracks allowed by domestic life: 'I rose early and made such arrangements with regard to my children and family affairs that I had time to write afterwards, not however without many interruptions. A man can always command his time under the plea of business, a woman has no such excuse.'

Mary's version of *Celestial Mechanics* was called *Mechanism of the Heavens*, because it was not a literal translation but an expanded version of Laplace's first two volumes. It was thus a stand-alone exposition of the mathematical language of gravity with specific application to astronomy, in order to make sense of how the solar system maintained its seemingly eternal and harmonious motion. No-one knew at the time, but Nathaniel Bowditch, the self-taught American mathematician, was currently making a literal English translation of all five volumes of Laplace's masterpiece. He also added explanatory footnotes, because, as he noted wryly, 'Whenever I meet in Laplace with the words, "Thus it plainly appears", I am sure that hours, perhaps days, of hard study will alone enable me to discover *how* it plainly appears!'

Mary no doubt agreed with this sentiment, because she, too, filled in many of the mathematical steps that Laplace had assumed were obvious. She also included many additional topics that helped give readers a fuller understanding of the basics of gravitational theory. Even so, when she submitted her manuscript to Brougham, three years after she had begun her secret project, he found it too complex for his intended audience, and he decided not to publish it after all. It was quite a betrayal of his initial agreement, and Mary must have been devastated.

William Somerville was determined his wife's efforts would not be wasted, and he approached London-based Scottish publisher John Murray. Murray and his father had built up a successful publishing house with a wide and innovative list, including the bestselling Byron, several of Jane Austen's novels, the bestselling *Domestic Cookery* by Mrs Rundell, and *De l'Allemagne* (*On Germany*) by Madame de Staël. A moderate revolutionary, prolific writer, thinker and salonnière, De Staël had left Paris during the Terror but she later fell out with Napoleon and went into exile again. During this time she met great German writers like Schiller and Goethe, but after she finished *De l'Allemagne* in 1810, Napoleon prevented it from being published in Paris because he deemed a book in support of German ideas and Romanticism to be 'un-French' (and therefore a threat to his empire: recall Napoleon's forces had defeated the Prussians in 1806). John Murray published De Staël's book in 1813, and it received much acclaim in Britain and Europe. He also published science books such as Charles Lyell's *Principles of Geology*. Lyell was a friend of the Somervilles, both of whom were keen amateur geologists. Geology was a new science at that time, and it was shaking up religious views about Creation because the geological record suggested the earth was far older than did a literal interpretation of the biblical account. In old age, Mary would remember the outcry over this question as being even more heated than the then current one over Darwin's theory.

Clearly, then, Murray was an innovative publisher, and he also hosted 'brilliant' dinner parties, as Mary later recalled, 'with all the poets and literary characters of the day, and Mr Murray himself was gentlemanly, full of information, and kept up the conversation with spirit'. Despite the fact that Mary was an unknown female mathematician, Murray was impressed that John Herschel had checked her manuscript and found very few errors. Herschel had graduated top of his mathematics class at Cambridge, and he had also taken over running his father's observatory, where he developed new photographic techniques for making astronomical observations – photography itself being a new invention at that time. Murray eventually agreed, on the strength of Herschel's report, to publish Mary's book in a small print run of 750 copies, because of the limited demand for such a technical work. (Recall the first print run of the *Principia* was around 500 copies.) She was forever grateful to him for this act of trust in her, and in this remarkable way, *Mechanism of the Heavens* was published at last, in 1831, when Mary was fifty years old.

It was received with great acclaim, because it broadened the accessibility of *Celestial Mechanics*, which was considered by contemporary scholars to be second only to the *Principia* in intellectual terms. There was also the occasional, inevitable, sexist, 'carping criticism': regarding one such criticism, Mary recalled, 'I was much annoyed, more so than I ought to have been for he showed that he was totally ignorant of the state of science.' But the general response was in line with that of Henry Warburton, who said no more than five men in Britain were capable of writing the book.

As I mentioned, Mary's book explained many of the mathematical definitions and techniques Laplace had taken for granted, and it also gave additional background material, such as the general theory of falling bodies derived from Galileo's experiments but expressed in terms of Newton's laws of motion, using Leibnizian calculus. Although it was not her conscious intention, this feature made Mary's book suitable for an advanced university textbook, and she was enormously flattered when professors Peacock and Whewell introduced *Mechanism of the Heavens* as a text for their higher astronomy classes at Cambridge. But Peacock, Herschel and Babbage had only recently begun teaching Leibnizian calculus in their mathematics classes, so it was not from mere flattery that they suggested Mary's book was eminently suitable as a cutting-edge text on celestial mechanics. (In fact, it would remain the standard text for the next century.) Although she did not say so, Mary must have felt an extreme sense of vindication at this turn of events, because, as a woman, she was not allowed to attend university herself: Girton College, Cambridge's first residential college for women, was not established until 1869, when she was eighty-eight years old. Back in 1831, she must have been thrilled to know that many reviewers commented on how extraordinary it was that a self-taught woman wrote a book that few men could understand.

The French, too, welcomed *Mechanism of the Heavens*. The Paris Academy of Sciences asked Biot to prepare a review to be read to the Academy and published in the *Journal des Savants*, and he wrote Mary what a pleasure this duty had been for him. He also expressed his long held admiration for her unique combination of charm and rigorous knowledge, the latter something that 'we men have been foolish enough to believe is our exclusive share'. He added that it was amusing to see even the 'most grave and least gallant' members of the Academy pressuring him to hurry up with his report: 'in a word', he told Mary, 'your book was a complete intellectual conquest here'.

Although most Britons had never heard of Laplace, the buzz around Mary's achievement engendered a spirit of national pride throughout the country. She was elected an honorary member of the Royal Astronomical Society – at the same time as Caroline Herschel, a fact that impressed Mary almost as much as the election itself. Many other honorary memberships followed, and although the Royal Society did not yet admit women, it commissioned a sculptor to create a marble bust of her, to be placed in the Society's Great Hall. Her female friends in particular were delighted with this singular honour. Lady Herschel, whose husband John had recently been knighted, wrote Mary to congratulate her on this 'richly deserved' honour, adding, 'I propose that we poor women, whom you have left so far in the background, shall raise a monument also, to shew our sincere love for you – and [to the fact] that you have not abjured your sex, while soaring far above it . . .' Similarly, on hearing the news of the Royal Society's decision, Mary's old friend Mrs Kater wrote her, 'I cannot tell you the delight with which we heard [of it].'

# *Chapter 15*

## *MECHANISM OF THE HEAVENS*

Like Émilie, Mary had not undertaken her book in the hope of financial reward, but rather for the glory of science, and for the personal challenge. *Mechanism of the Heavens* had been a massive undertaking: 610 dense pages, plus a seventy-page 'preliminary dissertation' providing an accessible, non-mathematical overview, analogous to the first part of Émilie's commentary. An illuminating contemporary response to the dissertation, and to Mary's literary ability, can be seen in a letter from her friend, the novelist and educational theorist Maria Edgeworth:

> [After reading it], I was long in the state of the boa constrictor after a full meal . . . My mind was so distended by the magnitude, the immensity, of what you put into it! [ . . . Because I am ignorant in scientific matters, I cannot judge what I have admired.] I can only assure you that you have given me a great deal of pleasure; that you have enlarged my conception of the sublimity of the universe, beyond any ideas I had ever before been able to form.

She went on to praise the simplicity of Mary's writing style, which 'particularly suits the scientific sublime . . . You trust [in] the natural interest of your subject, [in] the importance of the facts, the beauty of the whole . . .' She also listed some of the concepts in Mary's dissertation that had intrigued or amused her, including 'the moderate-sized man who would weigh two tons at the surface of the sun – and who would weigh only a few pounds at the surface of the [smaller] planets.' She concluded by referring to Mary's description of the propagation of sound: sound needs a medium like air to carry it from its source to our ears, and Mary wrote that if humans were able to travel beyond the earth's atmosphere and out into the vacuum of space, they would discover that 'the noise of the tempest ceases and the thunder is heard no more in those boundless regions, where the heavenly bodies accomplish their [orbits] in eternal and sublime silence'. Maria Edgeworth relished this passage, saying, 'It is a beautiful sentence as well as a sublime idea.' Then she added, 'Excuse me in my trade of sentence-monger,' but, of course, this very trade made her opinion of Mary's writing all the more valuable.

The preliminary dissertation may have been difficult for non-specialists to digest, but, as indicated earlier, the main body of *Mechanism of the Heavens* transcended the understanding of all but those with training in higher mathematics. To give an idea of the content of the book – and to show where scientific progress had been made since Émilie's time, and where it had not – the next few pages will explore Mary's treatment of several topics already mentioned in connection with Émilie's work, notably the brachistochrone, *vis viva* (or 'living force') and the stability of the solar system.

Recall that the brachistochrone refers to the 'path of quickest descent' for an object sliding down a smooth surface under the influence of gravity. In her *Fundamentals of Physics*, Émilie mentioned that Bernouilli and Leibniz had solved the problem by adapting Fermat's insight – namely, that a refracted light ray 'chooses' the quickest (rather than the shortest) route between two points in two different media. But she had referred the reader elsewhere for the mathematics involved, stating only that the required profile of the curved surface is that of a cycloid. Laplace had not discussed the brachistochrone, but Mary included it in her more extensive and accessible discussion of the way mathematicians are able to describe the various types of motion influenced by gravity.

She did not give the history of the brachistochrone problem, focusing instead on its mathematical solution, using the advanced technique of 'calculus of variations'. But back when Bernouilli had first posed his challenge, his older brother, Jacques, had submitted a solution that approached the problem via a rudimentary form of this method. Newton's solution, too, had contained the basic underlying concept of 'calculus of variations', because the idea is to choose an arbitrary curve, and then to show how to vary it bit by bit in order to obtain the precise curve that minimises some given quantity – in this case, the time of descent of the sliding particle. Euler and Lagrange developed the calculus of variations into a rigorous algebraic method (and to give the flavour of it, the Appendix gives a brief outline of how ordinary calculus can be used to find the minimum value of mathematical functions).

The calculus of variations incorporates – and, historically speaking, arises from – the important idea of 'least action', which Maupertuis and Leibniz independently pioneered by extending Fermat's approach to refraction. The principle of least action says the physical path taken by an object moving under given physical conditions can be found mathematically by minimising the object's 'action'. The mathematical method is the same as that used in the calculus of variations for minimising time in the brachistochrone problem, but minimising the 'action' refers specifically to economising on some 'active' quality of the motion, like its velocity, rather than simply minimising time or distance. Mary described the amazing way in which this principle can produce, *purely mathematically*, the basic Newtonian 'laws of motion'. These are the laws or axioms in which Newton defined the relationship between force, mass and motion, and it is these laws that are generally used to determine an object's path under a given force. But it turned out that these same laws follow as *mathematical consequences* of the principle of least action. (In the Appendix I have shown a simple example that Mary used to illustrate how this principle leads to Newton's 'first law of motion'.)

In other words, the principle of least action seemed to suggest that nature really does work in the most economical way possible, by 'choosing' the actual path of an object, from many theoretically possible paths, by minimising (or economising on) some aspect of the object's motion. In formulating their fledgling versions of this principle, Leibniz and Maupertuis had been deliberately guided by metaphysics: they believed some sort of supreme intelligence had designed the universe so as not to waste 'action'. Laplace did not go in for that kind of thing; he saw the principle of least action simply as a mathematical construct. Nevertheless, it had a vital role to play in his work. This is because French theoretical physics had become an almost purely mathematical subject, pared not only of metaphysics but also of intuitive physical analogies and concepts – including the intuitive geometrical concepts and diagrams of the *Principia*. In such a thoroughly abstract approach, it was necessary to find a way of identifying which of many theoretical mathematical laws can actually exist in nature, and in *Celestial Mechanics*, Laplace accepted only those laws that arise from the principle of least action, and also the 'principle of living forces'. In *Mechanism of the Heavens*, Mary retained this mathematical, non-metaphysical approach. (Personally, she believed the mathematical laws of nature were, indeed, evidence of design by God, just as Newton did, but, like him, she knew such beliefs had no place in science.)

Laplace's mention of the 'principle of living forces' indicates that some progress had been made on this topic since Émilie shot to fame for taking on the head of the Parisian Academy of Sciences: recall that in her famous dispute with Mairan, the debate had been over the very existence of *vis viva*. However, the idea was still not universally accepted even in Mary's time: in 1808, for example, her early mentor, Playfair, had published an article refuting her friend Wollaston's support for the concept. (Perhaps that is why Playfair criticised Émilie for changing her mind on the subject, in his note on her in his 1818 encyclopedia overview of the history of mathematics and physics: recall he accused her of not having 'exchanged the caprice of fashion for the austerity of science'.) Laplace and Mary did accept the existence of *vis viva* and, following Leibniz, they believed it was 'conserved' in some sense.

The fledgling idea of 'conservation of energy' was at the heart of the *vis viva* debate. Descartes, Huygens and others had noted the conservation of what we now call momentum in many kinds of collisions – such as when two billiard balls of the same mass directly approach each other with equal speeds and then, after colliding, they rebound with equal speeds in the opposite directions. (In such a collision, the masses and speeds are unchanged – but velocity includes the direction of motion as well as the speed, and momentum is proportional to velocity, not speed. In this case, however, the balls simply 'swap' momentum after the collision, so the *sum* of the momenta of the two balls is unchanged by the collision, which means the total momentum is 'conserved' throughout the process. The sum of the kinetic energies of the two balls is also 'conserved', as both Huygens and Leibniz realised.) However, in analysing more complicated motion than simple collisions, most notably falling motion, Leibniz discovered a brand new law that appeared to conserve 'living force' but did not conserve momentum. Consequently, he believed his new law was the 'true' *universal* conservation law.

Leibniz had deliberately searched for a quantity that would always remain the same, consistent with the pre-established universal harmony at the core of his monad philosophy. But his approach to *vis viva* was practical as well as metaphysical. I mentioned that in her *Fundamentals of Physics*, Émilie had described his argument about comparing the 'effort' of walking certain distances at certain speeds, but he also applied the same idea to the theory of machines, specifically those in which it is necessary to know how much effort is needed to lift a heavy object to a given height. Building on Galileo's analysis, he argued the required effort was equal to the 'living force' the object would gain if it were let fall from the same height.

In other words, effort is expended in raising an object from ground level to a certain height, but once it has attained the given height, the object has the 'potential' to acquire kinetic energy when it is let go; this 'potential energy' is equal to the kinetic energy the object will have acquired by the time it falls back to the ground. Leibniz also claimed that at any point *during* the fall, the *sum* of the object's 'living force' and the quantity we now call 'potential energy' is constant. He did not use this latter term but, following Galileo, he had a formula for it: gravitational force (or 'weight') times the object's height above the ground. This is the basis of the more general 'principle of living forces' that Laplace used, which, in modern language, says that at any point during motion caused by a force like gravity, the total sum of the kinetic and potential energies is constant or 'conserved'. This is called the conservation of 'mechanical' energy.

Leibniz (and Émilie) believed Newton had not considered this concept. The Appendix gives a fuller explanation of their reasons, but it seems they were misled partly because Newton never used the term 'living force': after all, the quantity defined by $mv^2$ is not a force but refers to what we now call 'kinetic energy'. However, Newton had indeed addressed the issue: in the *Principia*, he, too, spoke about the effort required in raising heavy objects, and he also spoke of 'overcoming any other given resistance by a given force'. Today we call this type of effort the 'work' done by the force in moving the object, and its formula is 'work = force × distance moved'. (In calculus, this can be expressed as the integral of force with respect to distance.) Newton gave an equivalent formula (and, as mentioned, Leibniz had a similar formula for the case of raising heavy objects against the force of gravity). But Newton's discussion was confined to a paragraph whose significance was generally overlooked for the next two centuries: until then, there was no suitable language to translate it into meaningful concepts like work and energy.

The *Principia* also contained three equally overlooked mathematical propositions, which, in hindsight, give a far more general and mathematically precise formulation of conservation of energy than anything written by Leibniz himself. The Appendix gives a little more detail, but the key point is that although Newton had the mathematics, and although Leibniz had the intuitive ideas, neither of them had a rigorous conceptual understanding of the notion of energy. After all, conservation of energy would not be experimentally verified until the late nineteenth century, which means that Laplace and Mary, too, were writing at a time when the idea was still hypothetical. Nevertheless, Continental mathematicians had independently rediscovered Newton's mathematical results (using Newton's laws of motion expressed in terms of Leibnizian calculus); consequently, Laplace's version of the 'principle of living forces' – which he expressed without reference to 'work' or 'energy' – is mathematically the same as the modern formulation of the conservation of mechanical energy (in which potential energy is expressed as a 'work integral', which gives the change in kinetic energy). For Laplace, this was a mathematical principle rather than a physical one; Émilie, too, had obtained the same modern equation, in her commentary on the *Principia*, but she did not connect it with work, conservation of energy or even 'living force'. Which goes to show how difficult it was to provide physical interpretations for the wealth of mathematics that flowed from Newton's laws.

It is interesting that Laplace still used Leibniz's term 'living force', which shows he did not mind the careless use of 'force' in this expression. Writing three decades after Laplace, Mary also used the term 'living force', which is further evidence that the idea of energy in physics was not yet clearly formed at that time. In analysing the situation of heavy objects being lifted against gravity, Mary concluded, 'So the impetus (or living force) is $mv^2$, [and it is] the true measure of the labour employed to raise the masses to the given heights.' Leibniz and Émilie had spoken of $mv^2$ as the true measure of *force*, (although Leibniz also used words like 'action' and, occasionally, 'kinetic action' and 'force of energy'), so Mary was closer to the mark with her use of the term 'labour'. She may have been influenced by her friend Thomas Young, who had discussed the term 'labour' in this context, and also the word 'energy' instead of 'living force' or *vis viva*. However, Mary's use of 'living force' illustrates the fact that the modern terms 'energy' and 'work' would not become widely accepted or understood until the 1850s, when William Thomson was one of the first to systematically utilise them.

Thomson – the future Lord Kelvin – pioneered the science of thermodynamics, building on the work of French engineer Sadi Carnot. Carnot might well have been the first to prove the conservation of energy had he not died of cholera in 1832, at the age of thirty-six. In the end, it was the experiments of James Joule and Hermann von Helmholtz, among others, which finally proved the law of conservation of energy in its fullest form, not just the 'principle of living forces' (or conservation of mechanical energy). Leibniz himself had realised the latter is not a truly universal law: for example, in a 'non-elastic' collision where two soft clay balls do not rebound after colliding but stick together and stop still, they end up with no 'living force' at all, and no potential energy either. Leibniz was forced to conclude that in such cases, the initial 'living force' of the balls as they approached each other before colliding was 'absorbed by the minute parts [of the colliding bodies and] is not absolutely lost from the universe', as he put it. Modern physicists would agree that the moving clay balls' kinetic energy was not 'lost', because it would have been converted into heat energy during the mutual deformation of the balls that occurred when they collided and stuck together. So Leibniz had come very close to articulating the idea that there are other types of energy, such as heat and sound energy, and that it is the total sum of *all* these types of energy that is conserved, not each particular form of energy. In 'Query 31' of *Opticks*, Newton, too, gave an intuitive statement of the conservation of all types of energy, but he did not spell it out as clearly as Leibniz had done.

In a series of experiments spanning more than twenty years – culminating in 1878, nearly fifty years after Mary published *Mechanism of the Heavens* – Joule established the equivalence of mechanical energy, electrical energy and heat. (A brief description of his ingenious experiment is given in the Appendix). Furthermore, Joule's and others' experiments showed that at any point in the experimental process, one type of energy might be converted into another – as when kinetic energy is converted into heat energy – but the *total* amount of energy is conserved. Joule's experiments also confirmed Benjamin Thompson's earlier experiment which suggested that heat is not a specific substance as had been assumed in Émilie's time, but rather a form of energy that is transferred between adjacent objects or media whenever there is a temperature difference between them.

If heat were a unique type of material substance, it would be conserved all by itself, like mass; instead, it flows from one medium to another when work is done to produce heat, just as potential energy changes into kinetic energy because of the 'work' done by gravity during falling motion.

In the twentieth century, conservation of energy and mass would be combined via Einstein's $E = mc^2$, which says that, in a reference frame in which an object is at rest, its 'rest energy' is proportional to its intrinsic mass or 'rest mass'. The proportionality factor is a matter of units of measurement, so this equation can be considered a definition: a particle's 'rest energy' is, by definition, its 'rest' mass. The powerful energy in matter at rest is normally locked away within its atoms – in the binding energy that holds its atoms and nuclei together, and the 'rest mass' energy of its elementary particles. This internal energy (or rather, a small fraction of it) is released most spectacularly in nuclear fission (which is currently used in nuclear reactors for electricity generation), and nuclear fusion, which is how the sun produces its energy: for example, it takes about 2 kilograms of the sun's mass to produce enough energy to illuminate the earth for one second. Conversely – and recalling Émilie's and Voltaire's essays on fire – heated metals gain *energy*, because as they are heated their molecules move more quickly and so their kinetic energy increases; but it is only in nuclear reactions that energy can be converted into a detectable gain in 'rest' mass (or weight), courtesy of $E = mc^2$. No wonder Voltaire could not come to a conclusive result in his experiments! (Nor could Benjamin Thompson: in addition to his experiments on the heat produced by friction, he had tried to detect a gain in weight when water and other substances were heated.) In such everyday situations, it is more practical to speak separately of the conservation of mass, and the conservation of energy.

Modern physicists also speak of the conservation of momentum. Leibniz pioneered the concept of conservation of energy, but he was wrong in assuming it was more fundamental or more universal than the law of conservation of momentum: kinetic energy and momentum can each be derived from Newton's law of force, and both these quantities are conserved in appropriate contexts. (The Appendix gives a brief illustration.)

Because the concept of energy was not fully understood for several decades after Mary wrote *Mechanism of the Heavens*, there had been relatively little development of this topic since Émilie's time. But things were very different when it came to a tantalising question left hanging by Newton in 'Query 31' at the end of *Opticks* – namely, would all the various planets and moons eventually distort each other's orbits so much that the harmony of the solar system would be irrevocably lost, unless God intervened?

One of the things that had attracted Émilie to the *vis viva* debate was that she believed Leibniz had used the conservation of 'living force' to answer this question scientifically, without the need to invoke divine intervention. In 'Query 31', Newton had casually commented that the orbits of the planets contained 'some inconsiderable irregularities, [which] may have risen from the mutual actions of comets and planets upon one another, and which will be apt to increase, till this System wants a Reformation'. Leibniz had mocked this assertion in a letter to Newton's disciple Samuel Clarke, whom Voltaire had met: 'Sir Isaac Newton and his followers have a very odd opinion concerning the work of God. According to their doctrine, God Almighty needs to *wind up* his watch from time to time; otherwise, it would cease to move. He had not, it seems, sufficient foresight to make it a perpetual motion.' The bitter edge is no doubt a reaction to the then-current priority dispute over calculus, but Leibniz clearly did not understand the import of Newton's comment. Recall that finding the moon's orbit is a complicated 'three-body problem' involving the gravitational tug-of-war between the moon, the earth and the sun. As for similar perturbations in the solar system as a whole, these include not only factors such as the influence of comets as they come close to various planets, but also the fact that when the huge planets Jupiter and Saturn come closest to each other during their revolution around the sun, their gravity is strong enough to measurably perturb each other's orbits (and those of other nearby bodies). Newton assumed such disturbances would keep multiplying, making the system inherently unstable, although he believed it 'will continue by the laws [of gravity] for many Ages' before it needed a 'Reformation'.

Leibniz believed his own idea of the conservation of 'living force' was enough to prove – on metaphysical grounds alone – that the solar system would keep going forever, without needing divine intervention. Conservation of energy does indeed apply to planetary motion, but Laplace showed it takes more than energy considerations alone, and certainly more than Leibniz's 'principle of living forces', to guarantee the stability of the solar system.

Recall that in order to find a mathematical description of the moon's orbit, Newton had used a 'perturbation method', in which the idea is to mathematically tweak the theoretical, 'two-body' orbit by a small amount, and then to keep on repeating the process in order to get a better and better fit with the real orbit, based on astronomical observations. Recall, too, that Clairaut had improved Newton's method to such an extent that he had used it to predict the 1759 return of Halley's comet to within one month of its actual appearance; an error of one month over the comet's approximately seventy-six-year orbit is quite small, and Clairaut's prediction had been hailed as a great achievement for Newtonian mathematical physics. But Laplace and his colleagues went much further, and developed perturbation theory sufficiently to apply it to the whole solar system, a topic that fills up much of *Celestial Mechanics* and *Mechanism of the Heavens*, and which requires an awe-inspiring amount of sophisticated, challenging mathematics.

Suffice to say, as Mary noted in her overview at the beginning of her book, that Lagrange and Laplace made the extraordinary discovery that the mutual perturbations of the various planets appear to be periodic. That is, they increase and then decrease over the aeons, but always within such narrow limits that the system remains stable. You can imagine a 'stable oscillation' by thinking of a stationary pendulum with the bob hanging down in the vertical position: if you then 'perturb' the bob by pulling it to one side by a small amount, it will oscillate backwards and forwards, but it will never stray far from the original vertical position.

The reason for this periodicity is that for any two planets, their relative position with respect to each other changes as they circumnavigate the sun. In the case of Jupiter and Saturn, Mary explained the situation as follows. Imagine the case where the two planets are on the same side of the sun, and all three bodies are aligned in a straight line. When aligned like this, they are said to be 'in conjunction', and they are at their point of closest proximity to each other, which means they are perturbing each other's orbits by the maximum possible amount: when two planets are at their closest point, the distance between them is the least, so, by the inverse-square law, their mutual gravity is strongest. But Saturn is further from the sun than is Jupiter, so by the time Jupiter has made one revolution around the sun and has returned to the original 'point of conjunction', Saturn has made less than half a revolution. This means it is now approaching the point directly opposite the point of conjunction – so it is almost as far away from Jupiter as it can be, which means its influence on Jupiter is much less than it was in the position of conjunction. It will take approximately two Saturn years and five Jupiter orbits – a total of about fifty-nine earth years – before the planets are close to their original alignment, and the perturbations are at their strongest again – and so on, with the perturbations waxing and waning in this way.

This is only a rough sketch of the situation, because Saturn and Jupiter are also affected by other planets as they approach and recede from either planet. As Mary explained, it turns out that most of these smaller perturbations compensate each other and cancel out, and the dominant effect occurs when the two planets are closest to each other. In fact, observations showed that over time, Jupiter's orbit appeared to be expanding and Saturn's shrinking, so that the two planets were actually pulling each other closer and closer together. But here, in Mary's words, is the extraordinary thing:

> If the conjunction always happened in the same point of the orbit, this uncompensated [effect] would go on increasing till the periodic times

> and forms of the orbits were completely and permanently changed – a case that would actually take place if Jupiter accomplished exactly five revolutions in the time Saturn performed two. These revolutions are, however, not exactly commensurable; the points in which the conjunctions take place advance each time by [about eight degrees], so that the conjunctions do not happen exactly in the same points of the orbits till after a period of 850 years. [Adding in the other perturbations, the cycle is completed every 918 years.]

By which time, according to Laplace's exhaustive and exhausting calculations, the various discrepancies have 'completely compensated' each other, and a new cycle begins. In other words, Jupiter's orbit increases and Saturn's decreases, but then, mid-way through the 918-year cycle, the situation reverses and Jupiter's decreases while Saturn's increases – so that over the millennia, the two planets slowly oscillate mutually to and fro, rather than being continually drawn together and ultimately crashing into each other.

It was a profound victory for the theory of gravity, because Saturn and Jupiter are large enough that their mutual perturbations had long been noticed, but until now, it had seemed these perturbations could not be accounted for. Indeed, the discovery that for all the planets, the myriad deviations from the idealised elliptical orbits seem to cancel each other out – or to oscillate between very narrow limits so as not to disturb the overall stability of the solar system – was an astonishing and comforting one. It seemed, after all, that Newton's theory was sufficient to answer this vexed question, without any need for divine intervention.

It would take another half century and more to uncover the first significant limitation of Newton's theory. It had to do with the theoretical prediction of Mercury's perihelion (the point in its orbit that is closest to the sun). In a perfectly stable, fixed elliptical orbit, the perihelion should occur at the same point each orbit, but because of the various perturbing influences, the underlying planetary ellipses *themselves* are slightly perturbed, their axes slowly rotating. This means Mercury's perihelion point is not the same each Mercury 'year' (about 88 earth days) – rather, it 'advances' around the orbit by a few 'minutes' of arc each century. (Recall that, by analogy with time, an angular minute is one-sixtieth of a degree.) Compared with the newly measured value found from direct physical observation of this phenomenon, the Newtonian prediction was out by a tiny forty-three *seconds* of arc per century (where a second is a sixtieth of a minute). Tiny or not, this discrepancy showed that Newton's theory was not perfectly accurate.

A smaller (less accurate) discrepancy between theory and observation in this matter had been noticed since the mid-nineteenth century (although this was more than two decades after Mary published *Mechanism of the Heavens*); Urbain Le Verrier had conjectured it was due to perturbations in Mercury's orbit caused by an unknown planet, which he called Vulcan. After all, Le Verrier had been instrumental in the discovery of Neptune (in 1846), whose existence had been sought in order to account for unexplained anomalies in Uranus's orbit. But Vulcan was never found. By the early twentieth century, however, Einstein's theory of general relativity predicted the advance of Mercury's perihelion with astounding accuracy, thus accounting for that tiny forty-three-second difference between theory and direct observation. This amazing feat left Einstein feeling 'beside myself with ecstasy for days'. It also turns out that Newton's equations must be replaced by relativity theory for a more precise view of the universe as a whole. In this respect, Clairaut was right – in essence if not detail – in claiming the inverse-square law is not universally applicable. Nevertheless, Newton's theory is sufficient for most calculations within the solar system, and it is very much simpler to apply than Einstein's equations. However, like Mercury's perihelion, the GPS navigating system is an example close to home where relativity does give a noticeably better accuracy than Newton's theory.

As for the stability of the solar system, improvements in Laplace's mathematical methods, along with the development of computers, have given a new slant on the question. The first important new insight came from Henri Poincaré at the end of the nineteenth century: his study of three-body orbits led him to an extraordinary discovery that laid the foundations of the now-famous chaos theory. The essence of chaos theory in this context is not that the world reveals itself to be totally chaotic and beyond the control of mathematics; rather, it means that ordinarily useful equations can sometimes become extremely sensitive to the accuracy of the experimental data inserted into them – to the extent that they become unreliable in making long-term predictions. (In the case of Newtonian perturbation theory applied to celestial mechanics, 'long-term' is a hundred million years in advance – a long time from a human perspective but not from an astrophysical and geological one.) This situation arises because in order to use equations to predict the configuration of a system of bodies at any given time in the future, it is necessary to know the objects' locations and velocities on at least one particular occasion. This data needs to be measured physically, so its numerical values are limited by the accuracy of the measuring process.

In many applications in physics, mathematical predictions tend to have the same degree of accuracy as the initial data, and as long as these inaccuracies are small, the long-term predictions are still acceptably accurate. What Poincaré discovered, however, was that when Newton's equation was applied to three or more mutually interacting celestial bodies, tiny inaccuracies in the initial data affected future predictions far more than had ever been suspected: instead of a one per cent inaccuracy in the initial measurement showing up as a one per cent inaccuracy in a long-term prediction, the inaccuracy compounded exponentially as time went on. The sense of 'chaos' comes into it because two almost identical sets of experimental measurements can give dramatically different long-term predictions, even though both predictions were made with similar data and with the same equation.

In hindsight, the fact that Newton's laws can uncover 'chaotic' behaviour is not so surprising – rather, it is an example of the uncanny connection between mathematical and physical structures, because many physical systems are so sensititive to initial conditions that they do indeed behave unpredictably. Weather systems are a notorious example, while the individuality of snowflakes is a beautiful one: it is not possible to predict the final shape of a snowflake because two adjacent 'seeds' released at the same time will follow different paths as they fall through ever-changing air currents, and so they will experience slight differences in temperature and other factors that affect crystal formation. But even a 'simple' game of billiards or pool is unpredictable, because if you tried to replicate it, tiny differences in the angle of the cue and the initial position of the balls could result in very different trajectories, and these differences accumulate with each hit.

Although this kind of sensitivity refers to the impossibility of making long-term *predictions* rather than to chaos in the usual sense of the word, the two tend to go together in such complicated systems as the solar system. In this case, a key issue is 'resonance', the phenomenon Mary alluded to when talking of the conjunction of Jupiter and Saturn: if the two planets came closest to each other *near the same point* on their orbits every fifty-nine years, then they would pull each other closer and closer every time, until their orbits were significantly changed. However, because of the complexity of planetary perturbations, not all resonances produce chaos, and not all forms of chaos are catastrophic (although over time, tiny deviations could accumulate and change everything). For example, resonance between Saturn's moons Hyperion and Titan has given Hyperion a (currently) stable orbit but a physically chaotic axis, whose orientation cannot be predicted even a few weeks in advance. On the other hand, Mercury and Jupiter seem to be heading towards a potentially catastrophic resonance in the distant future, whereby it is *possible* that Mercury will be pulled off course and crash into Venus or the earth.

Such predictions are derived from many pairwise applications of Newton's inverse-square law, together with a small correction for relativistic effects. But the discovery of mathematical unpredictability means that all these long-term calculations refer to *possible* rather than *actual* planetary behaviour; by computing many possible scenarios via successive recalculations with new initial data, the *probability* of catastrophic instability can be estimated. On the basis of these analyses, many astronomers believe that the solar system is almost certainly unstable. Nevertheless, its basic structure has remained stable for several *billion* years, and current mathematical analysis suggests it is unlikely that any of the planets will collide or wander out of the solar system during the remaining few billion years of the sun's productive lifetime.

Despite all these new developments, the perturbation theory developed by Laplace and Lagrange (and extended by Le Verrier and others) still provides an accurate approximation of planetary motion in the 'short-term'. But the existence of chaos – mathematical and physical – shows there can be no definitive prediction about the ultimate fate of the solar system. Which means that for all Leibniz's scorn about 'God winding up his watch', and his confident claim that the 'principle of living forces' was sufficient to keep the solar system in perpetually harmonious motion, Newton gets the last word: not only is his theory of gravity phenomenally accurate for most practical applications within the solar system, but his hunch about the inherent instability of the system cannot be disproved.

## *Chapter 16*

# MARY'S SECOND BOOK: POPULAR SCIENCE IN THE NINETEENTH CENTURY

In the aftermath of *Mechanism of the Heavens* – the three years of hard work writing it, and then the excitement and anxiety surrounding its publication and reception – Mary was exhausted and ill; cholera was raging in London, and she suffered severe bilious attacks. William recommended a rest and a change of scene, so they decided to return to Paris, in the summer of 1832. On their first trip in 1817, their infant daughters Martha and Mary had been left in the care of relatives, but now the girls were young women of seventeen and fifteen, and Mary thought the trip would be advantageous both for her own health and for the girls' education.

Although Martha and Mary were accomplished in music and languages, especially French and German, neither showed any interest in mathematics and science. But they were immensely proud and supportive of their mother's work, as was their half-brother, Woronzow, who was now a twenty-seven year old London barrister. He was sufficiently interested in science that he would be elected a Fellow of the Royal Society in 1833, but his interest was that of an informed supporter and amateur rather than a key player.

Mary's Parisian friends were eager to entertain her and to congratulate her on *Mechanism of the Heavens*: like the English, they, too, were impressed by the fact that it was a book written by a woman that few men could read. Woronzow visited his family soon after they arrived in Paris, and wrote back home of the extraordinary welcome his mother had received: 'Her arrival was proclaimed in glowing terms in the newspapers, and the French people generally seem to have been vying one with another who should show her the most honour.' So much so that Martha privately referred to the feisty Madame Rumford as 'a nasty old hag', because, according to Martha, she seemed 'jealous of Mama's meeting attention from [so many] people'. (Madame Rumford was none other than the widow of Lavoisier, and she was also the widow of her second scientist husband, Benjamin Thompson. In contrast to Martha's harsh judgment, Mary herself recalled that Madame Rumford was kind but 'capricious'.)

The Somervilles also made new Parisian friends, including the Lafayette family, who had played a key role in the early days of the Revolution. General Lafayette had drafted the Revolution's *Declaration of the Rights of Man and of the Citizen*, which was based partly on the American Declaration of Independence, which, in turn, had been inspired by Locke. Mary was also thrilled to make some new scientific acquaintances, including André-Marie Ampère. His pioneering work on electromagnetism would be commemorated by the adoption of the now common term 'amp' (or ampere) as the unit in which electrical current is measured.

Another new acquaintance was the Duchesse de Broglie, Madame de Staël's daughter. De Staël had recently died and was much lamented in Paris, but her daughter had taken over the role of salonnière, and Mary recalled that her receptions were the most brilliant in Paris: 'Every person of distinction was there, French or foreign, generally four or five men to one woman.'

In addition to the locals, many visitors to Paris also wanted to make Mary's acquaintance, including Nathaniel Bowditch's son, Henry, who had heard from Madame Laplace that Mary was in town. Henry, a medical student, was proud to be known as 'the son of the American translator of Laplace', and wrote his father that Mrs Somerville received him very graciously. He described her as mild, domestic, amiable – with a 'very pleasant' modesty about her knowledge. He was also struck by the fact that although she possessed 'an extraordinary mind', she seemed 'conscious of how little man can know'. He also appreciated her 'pleasant, mild voice' and her 'Scotch accent'. When Henry Bowditch later visited London, the Somervilles introduced him to the scientific and medical fraternity there.

For all that she enjoyed meeting people, Mary needed deeper intellectual stimulation, and before she left London for Paris, she had already begun work on her second book, *On the Connexion of the Physical Sciences*. It was to be a more popular book than *Mechanism of the Heavens*, albeit a serious one, like Émilie and Voltaire's *Elements of Newton's Philosophy* and Émilie's *Fundamentals of Physics*. At a time when scientific knowledge was growing rapidly, Mary's theme of unity – of showing the underlying 'connexion' between different branches of science – was important and fairly novel. New concepts were continually being formed as new experiments took place, and Mary's goal was to show how the different branches of science were increasingly coming together. For instance, the emerging connections between the various forms of what we now call energy included the connection between colour and heat. In this context, Mary mentioned William Herschel's 1801 discovery of solar infrared radiation, which arose from the experiment in which he also showed that different coloured light has a different temperature (just as Émilie had suspected).

Similarly, Alessandro Volta's invention of the electric battery, in 1800, led to new studies of electricity, and Mary described the connections being made between electricity and heat, between electricity and chemistry (such as inside Volta's battery), and between electricity and magnetism. In fact, electromagnetism itself was a new discovery, and Mary described its fascinating experimental history. The story began in 1819, when Hans Oersted noticed that turning on an electric current caused a nearby magnetic compass needle to be deflected, just as if another magnet were placed nearby. It was as if the accelerating electric charges in the current were both electric and magnetic at the same time. In 1831, just a year before Mary began work on her new book, her friend Faraday completed the discovery of electromagnetism by showing the converse of Oersted's discovery, namely, that a moving magnet creates an electric force, as if out of thin air.

Most of these new connections would not be fully understood until much later in the century, but Mary's book showed the new direction in which the physical sciences were moving. Four decades later, James Clerk Maxwell – creator of the mathematical theory of electromagnetism – would emphasise Mary's prescience: he mentioned her *Connexion of the Physical Sciences* as one of those 'suggestive books, which put into definite, intelligible and communicable form, the guiding ideas that are already working in the minds of men of science . . . but which they cannot yet shape into a definite statement'.

Like the *Fundamentals* and the *Elements*, Mary's *Connexion of the Physical Sciences* was a serious attempt to explain sophisticated scientific concepts to a non-specialist but highly literate adult audience; it was not a book primarily for the 'ladies', like Algarotti's *Newtonianism for the Ladies*, or even Jane Marcet's *Conversations on Chemistry*. Like Algarotti and Fontenelle, Marcet had based her book – and also her *Conversations on Natural Philosophy* – around conversations between female pupils and their teacher. In Marcet's case, the pupils were two teenaged girls, and their teacher was Mrs B, but Marcet was not merely copying her predecessors in using a conversational style: in her preface, she explained that she herself had learned what she knew about chemistry from 'familiar conversation' with a friend (presumably her doctor husband), and that this would be a good way to help other women learn the subject. After all, she noted, women's education 'is seldom calculated to prepare their minds for abstract ideas, or scientific language'. Accordingly, Mrs B conducted a whole range of interesting experiments to teach her students about physics and chemistry – and some of her experiments are still providing inspiration for classroom teachers today.

For instance, to show that different materials radiate heat differently, Mrs B rendered the sides of a tin in different ways – one side having a shiny polished surface, another blackened with soot, and so on – and then, with the help of a lens to focus the heat, she measured the temperature at the surface of each side after the tin had been filled with hot water. Mrs B also gave many lessons on the properties of steam, in order that her students could appreciate the wonderful invention of the steam engine. (Marcet's son, François, would later work with Arago on developing a device for measuring the pressure of steam – Arago was interested in bringing the British-invented steam engine to France, because, like Jane and Mary, he was excited by the use of science to create technology to improve everyday life.)

*Conversations on Chemistry* is an outstanding attempt to give an appreciation of practical, hands-on, basic science, so much so that it had inspired a young, barely literate Michael Faraday to take up a scientific career. Like William Wallace, Mary's first mentor, Faraday had begun his working life as an apprentice bookbinder, and Mary recalled that he never forgot his roots: 'At the height of his fame, he always mentioned Mrs Marcet with deep reverence.'

Marcet's book was used as a textbook in girls' schools, but as Faraday's experience shows, it proved popular with adults, too. Mary's book was not designed to appeal to such a wide range of readers, but as it turned out, the success of *On the Connexion of the Physical Sciences* exceeded everyone's expectations. It was first published in February 1834, and it would go through eight more editions in Mary's lifetime, and a tenth edition in 1877. (It was also translated into Italian and German.) By 1836, it was John Murray's bestselling science book and one of his bestselling books in any genre: it sold nearly seven thousand copies in its first three editions, and would sell around fifteen thousand copies all up. For comparison, Jane Marcet's *Conversations on Chemistry* – which was first published in 1806 – would go through sixteen editions in Britain, selling about twenty thousand copies all up; it also sold an estimated one hundred and forty thousand copies in the US, but many of these were pirated editions. Mary's book, too, was pirated, much to her anger: she therefore signed Harriet Martineau's petition to the US Congress, demanding copyright protection of British works in America.

Maxwell's comment on the prescience of *Connexion of the Physical Sciences*, together with the use of *Mechanism of the Heavens* as a textbook at Cambridge, show that Mary's high-level scientific writing was considered extremely important: it helped shape the identity of the emerging, increasingly professionalised scientific community, and it helped educate young scientists at a time when Britain had fallen behind France and Germany in mathematical physics. Consequently, Mary received yet more honours, including a private audience with the Duchess of Kent and the young Princess Victoria, to whom she presented a copy of her book. (Incidentally, several years later, in 1838, Mary attended Victoria's coronation: in her memoir, she recalled that the new queen was then a 'pretty young girl' of nineteen, who, despite her youth, 'went through the imposing ceremony with all the dignity of a Queen'. Writing three decades after the event, she added that Victoria 'has endeared herself to the nation beyond what any sovereign ever did before'.) By far the most important recognition of Mary's work, however, was the award of a civil pension of two hundred pounds a year. This incredible honour was announced in 1835: the government gave pensions to important writers and scientists, and another recipient that year was Faraday.

Eighteen thirty-five was an auspicious year astronomically speaking, too, because of the mathematically anticipated return of Halley's comet. Recall that Clairaut had accurately predicted the first return in 1759, and had published Émilie's translation of *Principia* to mark the event. Now it was Mary's turn, and she later recalled, 'The return of Halley's comet, in 1835, exactly at the computed time, was a great astronomical event, as it was the first comet of long period clearly proved to belong to our system. I was asked by Mr John Murray to write an article about it for the *Quarterly Review*.' The comet's return was still seen as a remarkable vindication of the theory of gravity, and in a letter to Mary at this time, the astronomer W. H. Smyth wrote, 'How I wish someone would give us a life of Newton, with all the interesting documents that exist of his labours.' (Incidentally, in 1801, Gauss had done for asteroids what Halley had done for comets: his calculations enabled astronomers to confirm that a mysterious object named Ceres – first observed by Giuseppe Piazzi – was an orbiting asteroid, the first to be discovered.)

On the domestic front, Mary's pension came at a crucial time for the Somervilles: their cousin had just defaulted on a large loan for which William had been guarantor, an action he had taken precipitately, without discussing it with Mary. For all his emotional and practical support of her – his editing and copying of her manuscripts, his liaising with Murray about publication, his role as a conduit for advice from scientists, and his own willing advice – William's flaw was that he was too easygoing, too inclined to be generous without proper consideration, and without discussing his financial dealings with Mary. As a woman, she had no legal say in the family finances, although it was not through sexism that William kept her out of the loop; rather, it was because he hated thinking about problems. He always assumed everything would turn out all right – and usually it did, although in this case, life would have been very miserable without Mary's pension and the royalties from her books.

Woronzow's reaction to the news of Mary's pension was another example of the support her children always gave her. He wrote to her saying that Prime Minister Peel had shown 'sound judgment and upright generous feeling' by being the 'first to appreciate the splendid talents and the indefatigable industry and the deep research displayed in the wonderful works which have come from your pen', and to reward at last those who 'sacrifice their time, their health and their fortune in pursuit of knowledge [that is] beneficial to the human race'.

But her success did not turn Mary's head: she told her children of the response of a couple of disgruntled citizens, who were overheard saying the government had been too generous in giving Mrs Somerville a pension for writing a book that nobody could understand. As for fame itself, she wrote Woronzow, 'I am a kind of tame Lioness at present, and Martha says that even she and Mary are being Cubberised.' (If Mary was the tame lioness, then her daughters Martha and Mary were her cubs, and they were enjoying a gratuitous celebrity that the down-to-earth Martha satirised with her term 'Cubberised'.) As for her friends' opinion of the way she handled success, the following letter from Jane Marcet seems to echo the opinion of everyone who knew Mary: 'You receive great honours, my dear friend, but that which you confer on our sex is still greater, for with talents and acquirements of masculine magnitude you unite the most sensitive and retiring modesty of the female sex; indeed, I know not any woman, perhaps I might say any human being, who would support so much applause without feeling the weakness of vanity.'

*On the Connexion of the Physical Sciences* was a remarkable achievement, as Maxwell's aforementioned comment shows. Its great strength was Mary's ability to explain important scientific ideas, as well as the connections between them. In describing Newton's theory of gravity, she discussed similar topics to those in Émilie's books, such as the moon calculations and the 'superb' theorems with which Newton proved the theory of gravity. But new information meant she was able to go further than this, by discussing the stability of the solar system and related topics. She even pointed out such facts as that Kepler's law of equal areas does not hold exactly, because the planetary orbits are perturbed from true ellipses, and that by comparing the observed discrepancies from the ideal law, the perturbations in a given orbit can be calculated.

She noted, too, that because the earth is flattened, one of the key 'superb' theorems does not strictly apply – namely, the theorem proving that gravity acts as though all the mass of a spherical object is concentrated at the point at its centre. But she went on to show that, when viewed from a large distance such as that between the earth and the sun, the earth appears essentially spherical, which means the inverse-square law can still be applied, with remarkable accuracy, as if the earth acts as a point particle. (As for the 'superb' theorem about forces *inside* a sphere, Newton proved this is also true for a spheroid such as the flattened earth.)

Mary also explained how physicists use mathematics to 'measure' the relative mass of the sun – or any celestial body whose distance from the earth is known – with respect to the mass of the earth. This method uses Newton's laws, but it is easiest to see how it works by finding the mass of the earth itself. Equating Newton's laws of force and gravitational attraction gives a very simple (Year-nine-level) equation, $g = \frac{GM}{r^2}$, where $M$ is the mass of the earth, $g$ is the gravitational acceleration for bodies falling from just above the surface of the earth (small variations due to altitude and latitude can be ignored here, so that $g$ is considered to be constant), $r$ is the radius of the earth, and $G$ is known as 'the gravitational constant'. The important point here is that, apart from $M$, everything else in the equation was already known: Galileo had measured $g$, and a fairly good measure of $r$ had been known since Eratosthenes's two-thousand-year-old calculation; the gravitational constant $G$ was first accurately measured by Lord Cavendish in 1798. The astonishing thing is that although it is hard work deriving the equations in the first place, and making the physical measurements for $G$, $r$ and $g$, once you have these at hand, it takes only very simple high-school algebra to solve the equation for $M$, or, in other words, to 'weigh' the earth 'as if it were on a scale or balance', as Mary put it. When you substitute the measured values for $G$, $r$ and $g$, you find $M$ is about six billion trillion tonnes (that is, a six followed twenty-one zeroes). Émilie and Voltaire had made a similar point about the use of simple algebraic laws (notably Kepler's third law) to calculate the distance between Venus and the sun or between Jupiter and its moons – a point also noted in the *Connexion*.

Although Mary explained many fascinating details of Newton's theory and its developments, one of the most important features of her book was the fact that it also dealt with newer topics, like electromagnetism and the wave theory of light. She had learned this cutting-edge science directly from the leading British and French experimenters of the early nineteenth century: not only Young, Wollaston, Kater, Biot, Arago and Ampère, but also Faraday himself, who gave her valuable feedback on her chapters on electricity and magnetism. Sadly, Kater, Wollaston and Young had died before seeing how their protégée presented their discoveries to a general audience.

But part of the clarity of *On the Connexion of the Physical Sciences* was due to the fact that Mary also had some first-hand knowledge of the art of scientific experiment. Her laboratory equipment had been somewhat domestic: like Émilie, she used whatever came to hand. She had carried out her first experiment in 1825, and, even before she began work on *Mechanism of the Heavens*, she had published her results in the prestigious *Philosophical Transactions of the Royal Society*. Because women were not allowed to attend meetings of the Royal Society, however, the paper had to be presented by William Somerville, although it was reported in the journal as having been written 'by Mrs. M. Somerville' and 'communicated by W. Somerville MD, FRS, Feb. 2, 1826'.

Mary's experiment was on the possible connection between violet light and magnetism, or, as she put it, 'on the magnetising power of the more refrangible solar rays'. The underlying nature of light was still a mystery at that time, and even magnetism was little understood. Nevertheless, by the early nineteenth century, some physicists had begun to suspect there was a connection between the two phenomena. For instance, D.P. Morichini claimed to have magnetised a needle using violet light, but he had not been able to replicate his result. Mary had heard of his claim – he had unsuccessfully tried to demonstrate his experiment to Faraday, who had visited him in Rome – and she decided to design her own experiment to see if she could obtain a definitive result.

Using a steel sewing needle, a glass prism to produce different coloured light, and a large lens (which she borrowed from Wollaston) to concentrate the light, she carried out a meticulously documented series of experiments that impressed her colleagues (John) Herschel and Wollaston. Herschel wrote about it to Oersted in Copenhagen, saying she had shown that 'a needle half exposed for a few hours quietly to violet solar light acquires a permanent magnetism. The exposed end becomes a north pole.' In her paper, Mary also reported that blue and green 'rays' were less reliable than violet in magnetising the needle, and they required longer exposure to produce an effect. She also found that even with violet light, larger needles, such as a large bodkin, did not become magnetic, 'the mass perhaps being too great'. She even made measurements of the ambient temperature on different days in order to ascertain whether or not 'heat' was a factor in the magnetising process, and she found it had no effect.

Her paper was published or reported in several international journals, and for the next three years Mary's result was widely quoted as authoritative. But then other experimenters showed the magnetic effect was *not* due to the violet light but to the steel itself. In her paper, Mary had described the precautions she had taken to ensure there were no extraneous magnetic sources in the vicinity, but steel is naturally prone to magnetism, and can be magnetised by a magnetic field or by friction. Presumably, she had inadvertently magnetised her needle during the process of testing for magnetic effects, by repeatedly bringing it close to a permanent magnet or by friction. (Incidentally, in his essay on fire, Voltaire claimed he had deflected a magnetic compass needle using an intense beam of light focused with a magnifying glass, but it turned out that the magnetic effect was due to some nearby iron, not the light itself.) This is the way science goes: experiments must be repeated independently many times, to make sure the right causal connections have been made. But Mary was mortified at having published something untrue, and she did not mention her experiment in her new book.

However, writing nearly fifty years later, in his 1873 *Treatise on Electricity and Magnetism*, Maxwell was sufficiently impressed by Mary and her experiment that he took the trouble to explain the flaw in its underlying premise, a flaw that had escaped those of her contemporaries who had initially praised her work. The problem was a misunderstanding of the nature of magnetic polarity, which is different from the electric case. For instance, magnetic poles always seem to exist in pairs: each magnet has both a north and a south pole. By contrast, electrically charged particles have either a positive or a negative charge. Furthermore, explained Maxwell, the two different poles of a magnet 'do not differ as light does from darkness', so it is not possible to produce a 'north' pole in a needle simply by exposing it to light. He added, 'We might expect a better result if we cause circularly polarised light to fall on the needle, right-handed light falling on one end and left-handed on the other' – but even this analogy was faulty, because two such polarised rays 'do not neutralise each other, but produce a plane polarised ray'. (This refers to light waves vibrating in a single direction or plane, not to an intrinsic polar duality. Polaroid sunglasses are made of a film that is transparent only for light waves vibrating in a single plane, so they reduce glare by filtering out all the other waves.)

Maxwell went on to say that, after many unsuccessful attempts at finding a connection between light and magnetism, Faraday eventually discovered that a magnetic force could actually shift the *direction* of the plane of polarised light. It was the first definitive proof of a connection between two such apparently different phenomena as light and magnetism, and Maxwell would take it to its ultimate conclusion in his theory of electromagnetism. But my point here is that in searching for such a connection, Mary had been in the vanguard of experimenters, and her achievement was nothing to be ashamed of.

With the success of her first two books, she eventually overcame her embarrassment and began a new series of researches, in 1835. She mentioned the seeds of this experiment in her *Connexion*, when she discussed various experiments designed to uncover a link between heat and light, and between chemistry and light, as in the new science of photography. Her own experiment was designed to discover the effects of different coloured light on paper coated with silver chloride. Faraday was engaged in similar research, and the two of them exchanged letters about their progress.

She was pleased with her results: she found that 'rays' transmitted through transparent rock crystal had the greatest chemical effect, followed by those passed through violet and blue glass, with green glass giving the least. She sent her findings to Arago, who communicated them to the French Academy of Sciences, which published them in its journal, *Comptes Rendus*. She would later publish a third experimental paper, on the effect of different coloured light on different coloured organic matter (specifically, vegetable juices). So little was known about the effect of sunlight on plants that John Herschel considered these experiments to be so interesting and 'elegant' that he communicated them to London's Royal Society; her paper was published in 1845, in the *Philosophical Transactions of the Royal Society*.

In hindsight, the elementary nature of all three sets of Mary's experiments illustrates how little was known about the interconnections between light, colour, heat, electricity and magnetism in those days before the idea of energy was properly formulated, and before Maxwell discovered the true nature of electromagnetism. Nevertheless, although she did not establish significant new knowledge, Mary's personal involvement in designing original experiments added to her ability to discuss clearly the latest research of her peers. This is evident in reading her books: whether she is speaking about mathematics or experimental science, she writes with authority and ease. In his review of *On the Connexion of the Physical Sciences*, Whewell commented on the fact that she was a 'person of real science', in contrast to most popularisers. And Faraday certainly appreciated the way she described his own experiments. He wrote to her soon after her book appeared: 'I cannot resist saying what pleasure I feel in your approbation of my *Experimental Researches*.' (These were first published at around the same time as Mary's book, which shows how up to date it was.) Faraday continued: 'The approval of one [competent] judge is to me more stimulating than the applause of thousands that *cannot* understand the subject.'

Had he been alive, Thomas Young, too, would surely have valued Mary's endorsement of his experimental researches, and her support for his controversial wave theory of light. As the next chapter shows, the study of light had come a long way since Émilie and Voltaire wrote their essays on 'fire'.

## *Chapter 17*

# FINDING LIGHT WAVES: THE 'NEWTONIAN REVOLUTION' COMES OF AGE

Mary began her discussion of light by noting the glorious atmospheric effects at sunset. In this spirit of wonder, she added, 'It is impossible to trace the path of a sunbeam through our atmosphere without feeling a desire to know its nature, [and] by what power it traverses the immensity of space . . .' She went on to describe Newton's particle theory of light, but she claimed that Young's experiments had proved once and for all that light was transmitted as a wave, not a particle.

In 1801, Young had discovered that when a beam of light is shone onto an opaque screen with two pinholes in it, the two thin beams that emerge from the pinholes actually interact with each other: they produce an 'interference' pattern analogous to that produced by water ripples. You can see this pattern by dropping two pebbles into a pond. The circular ripples that spread out from each pebble appear as a series of circular 'peaks' and 'troughs', and when a ripple from one pebble meets a ripple from the other, the two ripples seem to 'jump over' each other, momentarily producing a higher peak and then continuing on as before.

In other words, when a peak (or crest) from one pebble meets a peak from the other, they momentarily combine to form a larger peak; similarly, their troughs combine to form a deeper trough. But something else happens in the process: as the circular ripples spread out, there are places where the peaks from one pebble meet the troughs from the other pebble, and, at these points, the waves cancel each other out. These flat areas are called 'nodal lines'. The pattern of crisscrossing amplified peaks and troughs, together with the flat nodal lines, is the 'interference' pattern produced as the waves interact with each other. (Nodal lines can be difficult to see simply by watching the intersecting ripples on a pond, but they can be observed in a more controlled situation such as a ripple tank, where the ripples are generated by regular mechanical vibrations.)

Young's two pinholes were like the two pebbles in the pond, and he obtained the very same type of interference pattern – a pattern made not of water ripples but of bright and dark lines. He explained this pattern by assuming the brighter lines corresponded to the places where two light waves combined their peaks and troughs, thus increasing the wave amplitude and presumably increasing the intensity or brightness of light. The dark nodal lines corresponded to places where the peaks and troughs cancelled and 'destroyed' the light. You can also see this effect if you hold two fingers close to your eyes, leaving a tiny gap between the fingers: when you look through the gap towards a light globe or a sunny window, you can see tiny dark lines punctuating the light coming through your fingers. The gap is the equivalent of a single slit rather than Young's two pinholes, and it illustrates the phenomenon of diffraction, the bending of light around the edges of the slit: the dark lines are produced when some of the light waves are bent around the aperture, thus becoming out of phase with other rays and cancelling them out.

In fact, the phenomenon of diffraction provided further evidence that a wave model could successfully account for the observed effects of light. Recall Newton had thought particles were the only way to explain the sharp shadows of everyday objects in bright sunshine: the particles supposedly hit the object and bounced off it, so there was no leakage of light into the shadow, unlike water ripples lapping behind a rock. Mary pointed out that things were a little more complicated than Newton had supposed, because the wavelength of light is incomparably smaller than the wavelengths of ripples in a stream, so the bending of light around an obstacle is only noticeable when the obstacle itself is tiny, such as the boundary of the tiny aperture between your fingers. Another example of the bending of light can be seen when the tiny beam that emerges from a pinhole is captured on a photographic film: the image produced has fuzzy edges and it is larger than the pinhole itself, which shows the light has 'leaked' into the shadow area behind the boundary of the hole.

Surprisingly, Newton and Huygens had actually known about the effects of what we now call diffraction, but they had not grasped its import. It was Young's French colleague, Fresnel, who first explained diffraction patterns in terms of a wave model of light: he combined Huygens's geometrical wave constructions and Young's interference theory. Mary was an early advocate of Young's theory, and in her *Connexion of the Physical Sciences* she expressed her support for the existence of light *waves* – as opposed to the Newtonian idea of light *particles* – by saying of those dark nodal lines, 'It is contrary to all our ideas of matter to suppose that two particles should annihilate one another under any circumstances whatever.' (Imagine the disbelief of twentieth-century quantum physicists when they discovered that 'anti-particles' seem to exist (albeit momentarily), and when a particle meets its anti-particle counterpart, the two kinds of matter do, indeed, annihilate each other! Their combined masses and kinetic energies are transformed into electromagnetic radiation, according to Einstein's $E = mc^2$. Examples are the 'anti-electron' (or positron), a particle of the same mass as an electron but with a positive rather than a negative electric charge, while an 'anti-proton' has the same mass as a proton – which is much larger than that of an electron – but with a negative rather than a positive charge.)

But that is hindsight, and Mary believed it made more sense to suppose there existed an invisible cosmic medium that permeated deep space, so that waves of light could travel from the sun to the earth. After all, surely light waves needed a medium in which to travel, just like water waves – and sound waves, too: audible sound impacts on our eardrums through variations in pressure, and these 'pressure waves' need a medium of transmission, like air or water – you cannot have pressure when there is nothing to compress. Newton was the first to give a mathematical analysis of sound waves (which Laplace improved upon), so they were well understood by the nineteenth century: recall that Mary's description of what happens when there is no medium to carry sound had captivated Maria Edgeworth.

The idea of the cosmic ether had been around for millennia: recall that in previous centuries it had been invoked as the stuff of the cosmic vortices that supposedly caused planetary motion. That concept had now been dropped, but the 'luminiferous' ether was being revisited, although it was evidently so subtle it was undetectable, apart from the fact that it transmitted light. But having to assume the existence of such a mysterious, seemingly unknowable medium was problematic for modern, Newtonian physicists, so adopting a wave model of light came at a price.

Nevertheless, Mary's book showed that Young had made a very persuasive case, and that he had provided more evidence for light waves than interference alone. In particular, she noted that Young's idea of a 'wavelength' for light gave a beautifully simple explanation of colour. A 'wavelength' is the distance between two successive wave peaks. Although Young did not know what a light wave was actually made of, he was able to calculate wavelengths from light's interference patterns by analysing them as if they were made of water ripples – and what he found was that each colour has a slightly different wavelength, with red having the longest and violet the shortest. Such a simple, physical explanation of the spectrum had eluded even Newton.

Young's calculations were remarkably accurate, but then, as Mary showed, they were based on measurements made by Newton himself. In fact, Newton's *Opticks* had inspired Young's wave theory as much as Huygens's geometric wave theory had done, because although Newton conceived of light *itself* in terms of material particles, he had recognised there was also a periodic or vibrational aspect to some of the *behaviours* of light. Not even Huygens had seen this: he had focused on geometrical wave constructions that beautifully illustrated the phenomena of reflection and refraction, but he specifically rejected the possibility of periodic light vibrations. Ironically, he did this precisely in order to avoid the possibility of two different light waves interfering with each other: after all, when you look at two ordinary beams of light – such as two intersecting torch beams – they do not seem to affect each other in this way.

Newton's ideas on this topic were undeveloped and intuitive, and they were expressed in bizarre verbal language that shows how even a mind as great as his struggled to interpret what he had observed. The Appendix gives more details, but the key point is that Newton had applied both his theoretical vibrational concept and his formidable experimental skill to the analysis of the coloured patterns in thin transparent media like soap bubbles. Everyone has marvelled at the beautiful colours in soap bubbles, but Newton was the first to make a meticulous experimental study of these kinds of patterns, measuring the distances between different coloured rings and the refraction of the various colours. As Mary said, 'The determination of these minute portions of time and space . . . do as much honour to the genius of Newton as the law of gravitation.' And it was Young's genius that led him to recognise that these coloured patterns were interference patterns, and to apply his theory of wave interference to Newton's data, and to make the first ever calculations of the wavelengths of light.

All in all, Young's theory was a triumph of experiment and logical deduction. As with any novel idea, however, it received its share of detractors. After all, more work was needed to flesh out the theory – to find what light waves were made of – and so scientists of the calibre of Biot and Laplace continued to support the particle theory of light. (In fact, in 1798, Laplace had combined the particle model of light with Newton's theory of gravity to give a mathematical description of what we now call a black hole. His little-known English contemporary John Michell had independently done the same thing a decade earlier, but no-one took the idea seriously until it turned up as a consequence of relativity theory in the twentieth century.) The Newtonian particle theory had currency until 1850, when Jean Foucault found experimentally that light travels more slowly through water than it does through air. Recall Voltaire had tried to explain refraction in terms of Newton's idea that light particles were pulled by a force of attraction into a denser medium like water, and that because of the acceleration caused by this force, the particle model predicted light would travel *faster* in water than in air.

Nevertheless, Newton had been the only one to entertain the idea that light behaved like both a wave and a particle. He had been flummoxed by this deduction, but it is well known that modern physicists believe light does have both a wave-like and a particle-like nature, particles of light being called 'photons'. As I mentioned in connection with Émilie's essay on fire, photons are not Newtonian particles, because they have no intrinsic mass; rather, a photon is a 'quantum' or 'packet' of light energy. Einstein posited this idea in order to explain the so-called photoelectric effect, in which light can stimulate the emission of electrons from certain materials. The experimentally observed behaviour of this effect cannot be fully explained with a wave model of light, but requires the idea that each emitted electron has gained its energy from an exchange with a discrete 'particle' or 'packet' of light energy.

But these photons also have a wave nature, as can be seen in modern versions of Young's pinhole experiment (namely, the aforementioned two-slit experiment), which can be applied to light that is so weak only one photon at a time passes through the screen. Amazingly, the effect over time of many such particles passing through two tiny slits in the screen is entirely different from the effect produced when only one slit is open. In the latter case, when the photons that pass through the slit are recorded on a photographic plate, they register essentially as a series of points of light; this is just what you would expect from particles. But when the second slit is open, there is an overall pattern to the arrangment of all these dots, namely, a series of dark and bright areas, exactly analogous to the interference pattern produced by waves. (Earlier, I described this unexpected arrangement of light and dark patches by analogy with a miraculous formation of several distinct piles of letters in a letterbox. The same 'interference pattern' is seen when material particles like electrons or neutrons are fired through a double slit.) In some collective sense, these particles or photons each pass through both slits at the same time, just as a wave would do; they do not simply pass through one slit or the other without interference, because if they did the pattern on the photographic plate would consist simply of two overlapping patches of light. Of course, any verbal description of what happens during the two-slit experiment should not be taken too literally, because interpreting many of the bizarre behaviours of photons and elementary particles is just that, an interpretation. What is certain, though, is that both a wave model and a particle model are currently needed to fully describe the known behaviour of light (and of subatomic particles).

Fortunately, at the everyday level rather than the subatomic, light does appear to behave just like a wave, a fact that Young had helped to establish. Mary boldly claimed, 'The existence of an ethereal fluid is now proved,' by which she meant that if light was a wave, then it needed a medium of vibration. That was as far as the story of light went in the early 1830s, and she moved on to discuss electricity and magnetism.

Mary noted that 'we are totally ignorant' of the cause of electricity, both the static electricity produced by friction and the moving electric 'current' produced via Volta's battery. Static electricity had been observed since ancient times: for instance, if a glass rod is rubbed with silk, the silk 'sticks' to the glass electrostatically, because it has a 'negative' charge while the glass has a 'positive' one. However, if two glass rods have been rubbed with silk in the same way, then they are both positively charged and they repel each other. These different types of electricity were labelled positive and negative by Benjamin Franklin although, as Mary said, no-one knew what caused electrostatic effects. (We now know that electrons are the source of so-called negative charge, while protons are positive, so that the glass becomes positively charged when the silk rubs off some of the electrons in the atoms of the glass. The silk then contains the extra electrons from the glass, so it is negatively charged.)

Nevertheless, continued Mary, 'Various circumstances render it more than probable that, like light and heat, [electricity] is a modification or vibration of that subtle ethereal medium, which, in a highly elastic state, pervades all space.' These 'circumstances' were rather tenuous – after all, electricity itself was then considered to be a fluid, and it was not clear whether there were two different types of fluid, positive and negative, or one type, so that positively charged objects had a 'redundancy' of electric fluid, and negatively charged objects had a 'defect' or deficit. Mary favoured the latter hypothesis, but she said all the electrical phenomena could be explained either way. Just as with Newton's and Young's deductions about light, it is fascinating to see, in Mary's descriptions, just how much information about electricity had been gained through the insightful interpretation of experimental observation, even though the underlying theoretical concepts were so rudimentary.

The most remarkable of all these painstakingly acquired electrical insights were those that led to the discovery of electromagnetism by Oersted and Faraday, between 1819 and 1831. For instance, Mary noted it had long been known by mariners that lightning affected their magnetic compasses, but the electrical nature of lightning was not definitively established until the work of Franklin in the mid-eighteenth century; when a build-up of static electricity begins to move in a discharging spark, like lightning, it behaves like a momentary current. And, as Oersted had now proven, changing electric currents behave like magnets, which means they can influence other magnets, such as a mariner's compass. (Incidentally, Franklin had assisted Jefferson in drafting the American Declaration of Independence.)

Mary's friend Wollaston suggested, and Faraday confirmed, that this new type of electrically induced magnetism did not act in the way ordinary static electricity and magnetism did. Two stationary electric charges – that is, two objects electrified with static electricity (or two electrons, or an electron and a proton) – attract or repel each other along the straight line between them. Similarly, the magnetic force between the poles of two tandem-aligned bar magnets is also directed along the straight line between them. Gravitational force, too, acts in a straight line between two objects such as the earth and the sun. By contrast, this new type of electrically induced magnetic force caused magnetic objects to move in a *rotatory* motion, not in a straight line. Faraday's discovery was the prototype of something that would revolutionise modern life: the electric motor.

He followed this momentous discovery by developing the prototype of the electric generator itself, in order to illustrate the other half of the electromagnetic puzzle – namely, the answer to the obvious question: if an electric current could act like a magnet, could a magnet act like an electric current? While the question was obvious, the answer was not, and it was nearly twelve years after Oersted's discovery that Faraday was finally able to demonstrate that if you move a magnet through a coil of wire, an electric current magically begins to flow, even though the wire is not connected to a battery or any other source of electricity. (The problem for modern electricity generation is the means by which a huge magnet is kept moving in order to create this electricity: traditionally, steam has been used, created from the heat of burning coal or, more recently, solar or nuclear energy. Hydroelectricity is generated using the force of falling water to turn the magnets. Solar cells, by contrast, use a kind of inverse photoelectric effect called the photovoltaic effect.)

At around the same time as Faraday was making his revolutionary discoveries, American physicist Joseph Henry also discovered the existence of electromagnetism. In addition, he had used Oersted's discovery to invent the electromagnet, in which a core of iron is wound around with wire; the wire coil is connected to a battery, and the current induces such a strong magnetism in the iron that it becomes a temporary magnet of enormous strength. Mary mentioned that Henry's electromagnet was able to attract and hold nearly a ton weight of metal. Henry's work on electromagnetism was not as famous as Faraday's, partly because of Faraday's prodigious experimental output, so when Henry read Mary's book, he wrote in his diary, 'Was pleased to find she made mention of my experiments in magnetism.' Consequently, when he made his first trip to London several years later, he was keen to meet her, and he obtained an introduction at one of Babbage's soirées.

These electromagnetic discoveries, and the new technologies that had begun to emerge almost immediately – including the telegraph, pioneered by Ampère – were all due to the genius of experimental physicists and engineers. But the mathematicians were about to enter the scene. With their help, the deep connection between electricity and magnetism would prove so profound it would irrevocably change modern life: it would lead to the wireless technology that has transformed the way we communicate with each other, and it would change the way we think about physical reality itself.

As Mary showed in her *Connexion of the Physical Sciences*, the mathematical story of electricity and magnetism effectively began in 1785, almost a hundred years after Newton had developed the mathematical theory of gravity. In that year, Charles Augustin de Coulomb made incredibly delicate experiments through which he discovered that the electric force between two charged objects obeys an inverse-square law, exactly analogous to the inverse-square law of gravity. If the two objects are charged with the same type of static electricity, the force is one of repulsion, while opposite types attract: the only difference between the two cases is the direction, which is denoted mathematically by a minus sign or a plus sign, respectively. An inverse-square law also holds for the magnetic attraction or repulsion between two (isolated) magnetic poles. It is a remarkable situation: electric, magnetic and gravitational forces all behave according to the same mathematical pattern.

In an even more amazing illustration of the ability of mathematical language to uncover hidden connections in nature, Joseph Priestley had guessed there was an inverse-square law for stationary electric charges almost twenty years before Coulomb – and he did not need to make a single new experiment to do it. Instead, he applied one of Newton's 'superb' theorems to an experiment already made by Franklin. Franklin had found that when he electrified an insulated metal can, all the charge ended up on the *outside* of the can. He knew this because when he attached a cork ball to a silk thread and lowered it into the can, the ball did not pick up any charge when it touched the bottom; but it did become charged when it was made to touch the outside of the can. This reminded Priestley of the theorem in which

Newton had used the inverse-square law of gravity to prove there is no gravitational effect on a material particle placed anywhere inside a hollow spherical shell. Only bodies placed *outside* the shell can feel any gravitational force from the matter that makes up the shell. 'May we not infer from this that electricity is subject to the same [inverse-square] laws [as] gravitation?' wrote the perceptive Priestley.

Mary did not mention this stunning argument – perhaps she had not heard of it. Or perhaps she knew it was not enough to establish the inverse-square law of static electricity without experimental confirmation like Coulomb's, just as Newton had used Galileo's and Kepler's experiments to derive his law of gravity. Nevertheless, Priestley had uncovered a beautiful example of the way the patterns in mathematical theorems and formulae can reflect underlying physical patterns. (Or, as some believe, perhaps mathematical patterns reflect our *perception* of underlying physical patterns: this was essentially the view of Immanuel Kant, in his famous 1781 treatise, *Critique of Pure Reason*, in which he suggested our brains are hardwired to perceive space and time in an orderly, mathematical way.)

Coulomb's law for stationary (or static) electric charges had been derived directly from experiment, but in the 1820s, Ampère and his colleagues began to use mathematics to try to gain a theoretical understanding of the new phenomenon of electromagnetism. By the early 1830s, however, Mary noted that Ampère's mathematical descriptions of the forces between electric currents and magnetic forces had not been able to explain all the observed phenomena, or answer all the questions, that arose from Oersted's initial discovery. And little mathematical work at all had been done on trying to describe Faraday's half of the electromagnetic puzzle. Mary's book reflects this situation, in which electromagnetism was almost entirely an experimental subject, for which no adequate mathematical theory had yet been developed.

Surprisingly, it was the mathematically 'illiterate' Faraday who took the first major step towards a theory of electromagnetism, beginning in the 1830s while Mary was writing her *Connexion of the Physical Sciences*. In trying to explain how a magnet communicates its force to the surrounding wire in order to induce a current in the wire, he suggested it created some kind of magnetic stress and strain in the ether that supposedly filled the space between the magnet and the wire. Similarly, he assumed the same type of 'force field' filled the ether around stationary magnets or electric charges. He deduced a pattern for these fields from careful experimental measurements of electric and magnetic intensities at different points in the surrounding space, but his idea did not attract much attention. After all, Faraday was self-taught and lacked mathematical skill: he was an experimentalist rather than a theoretician, and he was unable to use the sophisticated mathematical language of the theoretical physicists who were trying to describe electromagnetism in those early years – notably Mary's French colleagues, and also Gauss in Germany and George Green in Britain.

These theoreticians had been building up their mathematical descriptions by using the obvious similarities between electromagnetism and gravity: not only the inverse-square law, but also the idea of action-at-a-distance. After all, two nearby magnets, or two nearby electric charges, appear to attract or repel each other immediately, with no need for direct contact – just as the earth and the sun appear to affect each other at a distance. Newton's own Continental contemporaries may have been unable to accept the idea that gravity acted for millions of kilometres across empty space, with no apparent medium or mechanism of transmission, but Newton's mathematical framework had worked so well in explaining all the observed astronomical phenomena that subsequent generations, from Clairaut to Laplace, had developed Newton's mathematics into a sophisticated algebraic language. No wonder their mathematical descendants were now applying these techniques to electromagnetism, which also seemed to act at a distance.

Indeed, these modern Continental mathematical physicists saw themselves as so firmly 'Newtonian' that they argued against Faraday, saying electric and magnetic forces did not need an imaginary, inexplicable ether as a mediator, any more than the force of gravity did. But Faraday remembered that Newton himself had *not* believed gravity acted at a distance, and that he had focused on examining only the effects of gravity, not the way it propagated: in the face of mainstream criticism of his field idea, he noted wryly that 'Newton was no Newtonian'.

Mary supported Faraday's view. Like him, she favoured the idea that light, heat, electricity and magnetism were related. After all, Faraday showed that magnetism can affect the polarity of light, and he and Oersted had shown the interconnection between electricity and magnetism. Because Young had all but proved light behaved as a wave, and because waves propagate through space – they do not act at a distance – Mary agreed with Faraday that *all* these phenomena were probably manifestations of mechanical impacts or stresses in the ether. In *Mechanism of the Heavens*, she had helped bring Leibnizian calculus to Britain, but now her support of Faraday's 'ether field' foreshadowed an unexpected change of fortune for Continental mathematicians.

Mary would not become part of the circle of young men who contributed to the rise of nineteenth-century British mathematical physics. In 1838, four years after the first edition of *Connexion of the Physical Sciences* was published, the Somervilles moved to Italy, partly because it was cheaper, and partly because both William and Mary seemed to enjoy better health there than in London. Although Mary and her friends and colleagues would remain in touch through mail or visits, she was too distant from the mainstream to make friends among the younger scientists.

Back in Britain, one of the rising new scientific stars was the aforementioned William Thomson. In 1841, when he was only seventeen years old, he took the first step towards providing a mathematical framework for Faraday's concept that electric and magnetic forces are transmitted through the mediation of an ethereal 'field'. But the most important impact of Thomson's initial breakthrough was the inspiration it provided for the man who would ultimately eclipse not only Thomson but also everyone else: James Clerk Maxwell. Although

Mary seems not to have known of these young movers and shakers, they certainly knew of her: recall Maxwell mentioned one of her experimental papers in his *Treatise on Electricity and Magnetism*, and he praised her *Connexion of the Physical Sciences* – presumably he had studied it, along with *Mechanism of the Heavens*, when he was a student at Cambridge.

Maxwell was born in 1831, the year Faraday made his famous discovery of electromagnetism; by the time he graduated from university and began his own research in the 1850s, little theoretical progress had been made in the subject. But everything was about to change, because Maxwell's genius lay in his recognition that new mathematical language was needed to break the deadlock.

In the action-at-a-distance model, the effect of a force emanating from a source located at point A is assumed to *instantaneously* affect a particle at point B – no matter how far apart the two points are. Mathematically speaking, this means there is no need to worry about what happens in the space between points A and B, and integral calculus is suitable for this kind of analysis. Maxwell, on the other hand, was intuitively drawn to Faraday's field idea, so he chose mathematical tools that did take into account the surrounding space – namely, differential calculus, and the new concept of vector algebra, created largely by Irish mathematician William Rowan Hamilton. (Maxwell also made novel use of the principles of least action and conservation of energy.)

With his new mathematical tools, and after more than a dozen years of prodigious effort, Maxwell finally succeeded where others had failed: he managed to derive a set of mathematical equations that quantitatively described all the known experimental behaviour of electromagnetism, just as Newton's law of gravity was derived from the known behaviour of planetary motion. But that was only the beginning of the mathematical theory of electromagnetism. When Maxwell further analysed his descriptive equations, he found something that took his breath away.

What Maxwell had pulled out of his mathematical description of the effects of electromagnetism were two 'wave equations'. Such equations had been pioneered by Émilie's colleague D'Alembert, in order to model the vibrations made in a plucked violin string, so the fact that this type of equation turned up as an intrinsic part of his description led Maxwell to conclude that electromagnetism did, indeed, propagate as some sort of wave or vibration. In true Newtonian style, however, Maxwell's theory of electromagnetism did not stop with the fulfulment of its initial goal – that is, with the mathematical unification of electricity and magnetism, and the prediction that electromagnetic effects are propagated as waves through space. The truly extraordinary thing was that Maxwell's wave equations had been deduced from calculations derived solely from descriptions of electromagnetic phenomena – and yet these wave equations *also* had the key characteristics of light waves; in particular, they had the same speed. The significance of this is that no-one had known what light was actually made of, despite Young's brilliant experimental deductions; but now Maxwell had the answer: light and electromagnetism were one and the same thing. Or rather, light was a *type* of electromagnetic radiation: Newton and Young had shown there was a spectrum of visible colours making up light, each colour having a different wavelength, but Maxwell suggested visible light, infrared heat rays, ultraviolet rays, 'and other radiation if any', were part of a continuous *electromagnetic spectrum*.

If electromagnetism travelled as a wave, however, then surely it must need an ethereal medium of transmission, as Faraday had suggested? Maxwell was not so sure: he realised the ether was a purely hypothetical construct, an assumed causal mechanism to replace action-at-a-distance. Astoundingly, for all that Newton had insisted on avoiding untested causal assumptions in a theory, many of Maxwell's colleagues – including Thomson – had been trying to create mathematical models of the ether that would offer (separate) intelligible mechanisms for the propagation of light and electromagnetism: it was just like the old Cartesian/Leibnizian way of doing things! But Maxwell was Newton's true heir: although he spoke of electromagnetic waves, and although he believed in what he called the 'probable' existence of the ether, he did not *assume* it in his theory – he did not assume any mechanical properties of the ether at all. Rather, he used mathematical language that was better adapted to seeing what, if anything, was going on in the space between electromagnetic objects, in order to describe the very same observed electromagnetic phenomena that everyone else was trying to describe with 'action-at-a-distance' techniques. Consequently, Maxwell's waves arose not from causal assumptions but from known experimental facts, and from the structure of mathematics itself.

The theory of electromagnetism was published in 1865; the Newtonian paradigm underlying it meant that if physical electromagnetic waves could be detected experimentally, it would be a valid confirmation of the theory. But the breakthrough experiment did not arise until nearly twenty-five years later, when Heinrich Hertz discovered electromagnetic waves that had the wavelength of what we now call radio waves. His extraordinarily detailed experiment proved that the effects of an electrically generated electromagnetic vibration – he used an electric spark leaping back and forwards between two terminals – could travel through space *without the aid of wires*, and that it had all the wave characteristics already noted in experiments with light. Maxwell's theory was confirmed, and the wireless era was born.

Not that Maxwell's theory was fully accepted even then. It was one thing for Newton's theory of gravity to be applied to moons and planets that could be clearly seen in the sky, but Maxwell's waves were invisible to the naked eye, and he described them purely mathematically; moreover, his electromagnetic 'field' was a mathematical 'vector field' – it was not a mechanical entity like Faraday's or Thomson's ether. Consequently, for all that his colleagues believed themselves to be 'Newtonian', history repeated itself and, as in Newton's case, many scientists did not believe that Maxwell's equations described physical reality at all. They did not understand that the analogy between light waves and mechanical water or sound waves was just that, an analogy, a hypothesis, and that Maxwell's equations described a completely different type of wave – namely, an ebb and flow of electromagnetic intensity, with electric and magnetic 'fields' of force locked in an eternal rhythm in which a fluctuation in one type of force induces a change in the other, and vice versa. This self-perpetuating intertwined series of 'vibrations' explained how light manages to travel billions of kilometres through empty space – there was no need for the mechanical pushes and pulls of a material ether.

It would take two more generations before the mainstream physics community fully understood that Newton's paradigm had come of age in Maxwell's theory. In fact, Maxwell had transformed the Newtonian method into something even more beautiful and sophisticated: Newton had chosen language designed to minimise the possibility of criticism based not on his theory's effectiveness but on its radical methodology, so it is not entirely clear how much physical reality he ascribed to some of his theorems; Maxwell, on the other hand, was clear about the fact that mathematical language may be the closest we can ever come to perceiving the more subtle kinds of physical 'reality'.

*Chapter 18*

# MARY SOMERVILLE: A FORTUNATE LIFE

While Maxwell was developing his theory of electromagnetism in Britain, Mary was living in Italy with her husband and two daughters, Martha and Mary. The girls never married, perhaps because it was difficult for foreign girls with no fortune to meet eligible men in a country with different customs and social expectations. Or perhaps they were too independent to marry: they went sailing in their own boat (which their brother, Woronzow, had given them), they climbed mountains, travelled through Italy and France, helped their mother with her research, and joined in the social life of the intellectual elite. (As for Woronzow, he was a London barrister, as I mentioned earlier, and he had married in 1837 when he was thirty-two. When he was only nineteen or twenty years old, however, he had fathered an 'illegitimate' daughter, whom he secretly supported financially.)

Italy was then a group of city-states, including the papal states, the Austrian-occupied regions around Milan and Florence, and the kingdom of Naples; recall Émilie's daughter had been a lady-in-waiting to the Queen of Naples. Mary's letters to Woronzow show that during the war against the occupying Austrians in 1859, she and her daughters had provided bandages for wounded Tuscan soldiers, and she gave first-hand accounts of various events during this tumultous time – including the victory procession of the newly proclaimed king of Italy, Victor Emmanuel II, and the public outpouring of grief on the death of Count Cavour, a hero of the independence movement. In her memoir, Mary also described the immense admiration that she and many others felt for Giuseppe Garibaldi, who was another leader in the fight for the independence and unification of the country. (Mary noted that one female admirer was so inspired by Garibaldi she went so far as to collect the hairs from his comb.) France, too, was in a period of political upheaval, and Mary lamented the fact that, during the anarchy that accompanied the birth of the modern French ('third') republic in 1870, 'Laplace's house at Arcoeuil has been broken into, and his manuscripts thrown into the river, from which someone has fortunately rescued that of the *Mécanique Céleste*, which is in his own handwriting.'

During her 'retirement' in Italy, Mary wrote two more books: the first of these was *Physical Geography*, first published in 1848, with five more editions in her lifetime and a sixth shortly after. This book included discussion of geology – whose findings contradicted the biblical estimate of the age of the earth, as I mentioned earlier – and Mary recalled that its publication had precipitated an anti-geological sermon directed against her by a minister in York cathedral. She added that eventually, after quite some decades, the controversy died down, because 'facts are such stubborn things'. She also mentioned that *Physical Geography* was awarded the Royal Geographical Society's Victoria medal. It was 'an honour so unexpected, and so far beyond my merit, [that it] surprised and affected me more deeply than I can find words to express.' The Italians, too, recognised her work: she was already an honorary member of many Italian scientific societies and universities, but she was especially delighted to receive the inaugural gold medal of the Geographical Society of Florence.

Her final book was *On Molecular and Microscopic Sciences*, which was published in 1869, when she was eighty-eight years old. It was not reissued, partly because Mary was no longer mixing with leading experimentalists, and the book lacked the cutting-edge quality of her earlier works. The molecular nature of matter was still hypothetical in 1869, but dramatic new discoveries were about to revolutionise the subject, culminating in J.J. Thomson's discovery of the electron in 1895, and the development of quantum theory in the early twentieth century. But even without the benefit of hindsight, Mary herself realised *Molecular and Microscopic Sciences* was 'a great mistake, [which I] repent. Mathematics are the natural bent of my mind. If I had devoted myself exclusively to that study, I might probably have written something useful, as a new era had begun in that science.'

She was led to this realisation partly by an unexpected visit from Professor Pierce, a Harvard professor of mathematics and astronomy: he rekindled her interest in higher mathematics, which she had neglected while writing her popular books. Following Pierce's visit, she ordered the latest mathematical textbooks and studied them for several hours every morning. In the afternoons, she spent time with her daughters, or read literary rather than mathematical works: Shakespeare, Dante, 'and more modern light reading', as well as newspapers. In the evenings, 'I read a novel, but my tragic days are over; I prefer a cheerful conversational novel to the sentimental ones. I have recently been reading Walter Scott's novels again, and enjoyed the broad Scotch in them.'

In the same year, 1869, Mary wrote to philosopher and politician John Stuart Mill, on the recent publication of his book *On the Subjection of Women*. One of the great nineteenth-century philosophers on liberty, Mill was a major intellect – and, like Voltaire, he freely admitted he had had an intelligent woman by his side to help him sharpen his ideas, namely, Harriet Taylor. As a young man, he had been arrested for distributing birth control information in London's poorer districts, and, as a liberal Member of Parliament in the late 1860s, he campaigned for votes for women. Mary had been the first to sign his petition for women's suffrage. She wrote in her memoir, 'The British laws are adverse to women, and we are deeply indebted to Mr Stuart Mill for daring to show their iniquity and injustice.' (Inspired by the views of his late wife, Harriet, Mill pointed out that legally speaking, married women were in a position of slavery: they had no property rights – this would remain the case until the Married Women's Property Acts of 1870 and 1882 – and they could be physically beaten, compelled to have sex, and physically forced to return to abusive husbands. Moreover, legal separations were rarely granted, and women had no legal custody of their children.) Mary continued, 'The law in the United States is in some respects even worse, insulting [our] sex by granting suffrage to the newly-emancipated slaves, and refusing it to the most highly-educated women of the Republic.'

It would take another sixty years before all British women won the right to vote, in 1928. Back in 1869, *On the Subjection of Women* was causing heated controversy, and Mill responded gratefully to Mary's letter of support: like Faraday, he especially appreciated her approbation because he knew she understood his ideas. He described her as having 'rendered such inestimable service to the cause of women by affording in [your] own person so high an example of their intellectual capabilities'.

But Mary was an advocate for women as well as an example: she not only signed petitions, she was also a member of the General Committee for Woman Suffrage in London. She wrote, 'Age has not abated my zeal for the emancipation of my sex from the unreasonable prejudice too prevalent in Great Britain against a literary and scientific education for women.' She believed the French were 'more civilised in this respect', because 'they have taken the lead, and have given the first example in modern times' of the 'encouragement' of women's higher education – namely, Emma Chenu, who had recently graduated Master of Arts, with a diploma in mathematics, from the Paris Academy of Sciences. Mary added, 'A Russian lady has also taken a degree.' She did not name the 'lady' or her field of study, but by 1870 a number of young Russian women were studying medicine, chemistry or mathematics in Switzerland or Germany; the most famous of these women, then and now, is Sonia Kovalevskaia, the first woman to obtain a modern doctorate in mathematics, which she took from the University of Göttingen in 1874. Several years earlier, at the time Mary was writing, Sonia was studying at the university at Heidelberg, although women were not yet allowed to take formal examinations. The same situation prevailed at most universities, including Cambridge, despite the existence of Girton women's college: women could attend classes but they would not be granted full degrees at either Cambridge or Oxford until well into the twentieth century. Mary noted that she had signed a petition to London University requesting that degrees be granted to women, but it was rejected.

At eighty-eight, Mary was hard of hearing, and her arthritis meant she was having trouble holding her pen. She tired more easily than she used to do, but she counted herself extremely lucky to keep her intellectual vigour right up till the end of her life. As part of her newfound study of higher mathematics, she began to rework a 246-page mathematical manuscript on curves and surfaces, which she had written years earlier but had laid aside and forgotten (because of the pressure of bringing out a new edition of *On the Connexion of the Physical Sciences*). She took a keen interest in other scientific topics, too: for instance, in 1871, she was delighted when John Murray sent her a copy of Darwin's new book, *The Descent of Man and Selection in Relation to Sex*. Mary thought Darwin had presented his thesis (of evolution and natural selection) 'with great talent and profound research'. She added, 'In Mr Darwin's book, it is amusing to see how conscious the male birds are of their beauty.' But she was not amused at the vanity of women who, 'without remorse', 'allow the life of a pretty bird to be extinguished in order that they may deck themselves with its corpse'. (Unlike many of her contemporaries, Mary supported the theory of evolution, although she pointed out that Darwin himself acknowledged he had not explained the origin of what she called 'the first organic forms, the primordial types or varieties' from which evolution then proceeded – so she believed – according to some overarching natural law.) As for another 'remarkable' book Murray sent her – Edward Tylor's *Researches on the Early History of Mankind, and the Development of Civilisation* – she accepted his hypothesis that human culture had developed according to certain evolutionary laws, but she commented, 'Yet one cannot conceive human beings in a more degraded state than some of them are still; their women are treated worse than their dogs. Sad to say, no savages are more gross than the lowest ranks in England, or treat their wives with more cruelty.'

In her ninetieth year, she had begun fleshing out her memoir, which she continued to update over the next two years. She mentioned that William Somerville had died in 1860, when he was eighty-nine years old. It was not her style to describe her own grief, but her daughter Martha – who later edited her mother's recollections for publication – included an eloquent letter from Mary's friend Frances Power Cobbe: she expressed her own grief at the passing of 'the dear, kind old man, whose welcome so often touched and gratified me', and she wrote that William and Mary had represented 'the most beautiful instance of united old age. His love and pride in you, breaking out as it did at every instant when you happened to be absent, gives the measure of what his loss must be to your warm heart.'

Naturally enough, Mary's thoughts frequently turned towards her own death. She was sustained by her religious belief in an afterlife in which she would be reunited with her loved ones; nevertheless, she admitted she was afraid at the thought that 'my spirit must enter that new state of existence quite alone. We are told of the infinite glories of that state, and I believe in them, though it is incomprehensible to us; but as I do comprehend [. . .] the exquisite loveliness of the visible world, I confess I shall be sorry to leave it. I shall regret the sky, the sea, with all the changes in their beautiful colouring; the earth, with its verdure and flowers.' She was also sad at the thought of leaving the birds and animals that had given her such delight and companionship throughout her life.

Mary did not dwell on such thoughts, however. She knew she had had a fortunate life, although she had suffered her share of sorrow, especially the deaths of her three young children, her beloved husband, and her dear Woronzow, who had died suddenly at the age of sixty, in 1865. It was the very year Maxwell published his theory of electromagnetism, and Mary's deep depression over her son's death helps explain why she seems not to have taken much notice of this momentous scientific publication. After all, it remained 'unproven' for the next twenty-five years, and neither she nor her friend Faraday – nor Maxwell himself – would live to see the discovery of radio waves, or to witness the fact that from the twentieth century on, mathematical forces like those of electromagnetism and gravity would be conceived in terms of fields rather than action-at-a-distance.

In 1872, Mary described herself, at ninety-one years old, as 'extremely deaf', with her memory of ordinary events beginning to fail. Surprisingly, her mathematical memory still remained acute, and she continued to study mathematics for several hours a day; when she encountered a difficulty, she found that 'my old obstinacy remains, for if I do not succeed today, I will attack [the problem] again on the morrow'. But inevitably, her memoir had to end. On its last page, she noted that she still took a lively interest in current affairs, including science, although she regretted she would not live to see the results of the new scientific expeditions then being undertaken (including attempts to reach and study the North Polar Sea). But she regretted most of all that she would not live to see the suppression of 'the most atrocious system of slavery that ever disgraced humanity', as publicised by Livingston and Stanley, and which 'Sir [Henry] Bartle Frere has gone to suppress by order of the British Government'. (Frere would successfully sign a treaty with the king of Zanzibar, abolishing the slave trade that had existed between Zanzibar and Portugal.)

Then, ever the mariner's daughter, she added, 'The Blue Peter has been long flying at my foremast, and now that I am in my ninety-second year, I must soon expect the signal for sailing. It is a solemn voyage, but it does not disturb my tranquillity.' Indeed, Martha recalled that in recent years, her mother had often claimed that 'not even in the joyous spring of life had she been more truly happy'. After all, she had managed to combine a happy family life with intellectual challenge, and she had gained the respect of the international scientific establishment. Even her old age was made easier by the devoted care of her daughters.

On 28 November 1872, a month short of her ninety-second birthday, Mary was working on a mathematical paper on 'quaternions': these were Hamilton's invention, and they contained the vector algebra Maxwell had used in his 1865 theory of electromagnetism. The topic was still so new in 1872 that, back home in Scotland, Maxwell was exploring and developing the language of quaternions for his 1873 *Treatise on Electricity and Magnetism*. Maxwell's *Treatise* was the electromagnetic equivalent of Laplace's *Mécanique Céleste* – a compilation and mathematical analysis of all the experimental and theoretical investigations that had culminated in his own theory of electromagnetism. Mary would have loved it! But on the morning of 29 November she died in her sleep, after an extraordinary life well lived.

*Epilogue*

# DECLARING A POINT OF VIEW

During her life, and after her death, Mary was widely celebrated as the 'Queen of Science', and 'the most extraordinary woman of her time'. She had always been a vocal supporter of women's education, and she was immensely pleased with the news about Girton, the first residential women's college at Cambridge, which was established when she was eighty-eight years old. But she would have been overwhelmed to know that, in 1879, Oxford's new Somerville College for women would be named in her honour. And she would have been delighted that, by the early twentieth century, young women with a love of mathematics would no longer have quite the same struggle as she did.

Nevertheless, in most countries today, men still outnumber women in the mathematical sciences. I was intrigued that Mary noticed there was only about one woman to four or five men in the most prestigious Parisian salon of the day: it is a big step from a salon to a university, but I, too, have been in the habit of calculating the gender balance in various academic settings (perhaps it is natural for a mathematician to do so!). In the late 1980s, at my first international conference on general relativity, the ratio was about one to forty. Of course, the picture is not nearly so skewed in science as a whole, although Mary's figure of one in four or five still has a ring to it: for instance, in 2005, a Canadian study of women in astronomy showed that about one in five doctoral students were women, while a 2006 UNESCO study found that just over a quarter of the world's scientific researchers (in all disciplines) were women. Figures from my own department, the School of Mathematical Sciences at Monash University, show that in 2010, women made up only thirteen per cent of the academic staff – which is just below the average percentage of female science professors in Europe – but the postgraduate ratio was unusually high: three women to four men, although for honours undergraduate students, the figure was about one to three.

I have given further statistics in the Appendix, but perhaps I have already made the point that female heroines remain important for women in mathematics and physics. Fortunately, despite the relative lack of women in science, there are an increasing number of contemporary women who can act as role models for young women – for example, in 2008, Professor Penny Sackett, a physicist and astronomer, was appointed as Australia's Chief Scientist. But my conversations with colleagues and friends suggest that pioneers like Mary and Émilie still provide inspiration, not only for those working or studying in mathematics, but also for anyone who is interested in the history of ideas.

From my own point of view, Mary and Émilie have long been important in my life, because my path to becoming a mathematician has not been easy or straightforward. I have drawn comfort and inspiration from their courage and determination, and I am amazed at their achievements. They may not have proved any new theorems or created new theories, but few people ever make world-changing original discoveries. On the other hand, relatively few people today have such a deep knowledge across such a broad range of topics as did Mary and Émilie. Their struggle to acquire this knowledge was truly heroic.

In a way, the point of view of this book is that the history of scientific ideas is heroic, too. I use words like 'heroic' and 'seduced by logic' advisedly, but that does not mean I am uncritical of science. In fact, I once rejected it so completely that I dropped out of university and, inspired by the nineteenth-century writer Henry Thoreau, I 'went into the woods to live deliberately'. A handful of like-minded friends joined me. One of our disenchantments with mainstream life was the way it depended on science to create technologies that alienated people from nature, polluted the environment and enabled the development of horrific weapons. For these reasons, among others, living deliberately meant living simply, without electricity or running water or telephone.

Needless to say, I gradually softened my perspective on science, not just because of the physical difficulties of such a primitive life, but also because of the human need for intellectual diversity. My first concession to technology was a battery-powered radio: after a year of relative solitude, it was utterly thrilling to hear different voices, different conversations. For the first time, I experienced a *personal* appreciation of the profound importance of Maxwell's mathematical prediction of the existence of radio waves.

The next concession was the acquisition of a set of old army field telephones that could be connected by cable to each of the dwellings in the community. These cottages were about a kilometre apart across difficult terrain, so it was important to be able to keep in touch at the turn of a handle (this was long before the invention of mobile phones). These old-fashioned telephones were a science lesson in themselves: they did not need an outside grid because their power was obtained by manually turning a handle connected to a magnet, and the moving magnet created an electric current. Experiencing the benefits of electromagnetism in such a simple, direct way was exciting in itself, but it also illustrated the obvious fact that appropriate technology, rather than technology used profligately, is a good thing.

But it was the numinous aspect of the wilderness that had the most effect upon my change of heart about science, and about mathematics in particular, because, living so close to nature, the universe was always on my mind, and mathematics is the language of the universe. At night, the sky was so dark and the stars were so brilliant that, like Mary, I found it impossible not to wonder at the immensity of space; I yearned to understand its mysteries, and when I began to study again, I realised anew, like the passionate Émilie, just how calming it can be for an emotional person to abandon herself to the logic and certainty of mathematics, and to be seduced by its noetic beauty.

Both Mary and Émilie responded palpably to the intellectual elegance and economy of the mathematical language, and to the satisfying repetition of patterns in the laws of nature that give mathematics a beauty analogous to that of poetry and music. They were in awe of the power of Newton's one simple equation to describe and predict how physical things move both on earth and in space, and they would have been astounded at the way Einstein's equation of gravity 'contains', in its half dozen elegantly economical symbols, the structure of the universe itself. It is this aspect of mathematics that fascinates me, too, and which I have emphasised in this book.

As for the progress of the Newtonian/Maxwellian revolution, it took nearly half a century after Maxwell before mainstream mathematical physicists felt as comfortable visualising abstract mathematical structures like vector fields as they did with concrete mechanical models. Consequently, experimentalists kept devising various ways of trying to detect the ether – but none of these experiments yielded definitive, unqualified proof of its existence.

The centuries-old conundrum of the ether was finally resolved, at least for the moment, when the baton of mathematical physics passed back again to the Continent, specifically to Einstein, who created the next great theory in true Newtonian tradition. The 'special' part of the theory of relativity grew directly out of Maxwell's theory of electromagnetism, which had left dangling the question of the existence of the ether. Although his electromagnetic waves were not mechanical compressions or vibrations of a physical medium but were waves of electrical and magnetic intensity, the ether still had possible relevance as a fundamental frame of reference for measuring the speed of these waves (that is, the speed of light). By creating a mathematical framework in which physical laws like Maxwell's were formulated entirely relativistically – with no reference to any fundamental frame – Einstein formally did away with the need for the ether. This was possible because, according to Einstein, the speed of light is constant (in a given medium), the same in any frame of reference.

The 'general' part of Einstein's relativity theory is a 'field' theory of gravity, which was designed to extend Newton's theory, using a complex version of Maxwellian-style field language. In so doing, Einstein defined gravitational force in terms of a curving or warping of four-dimensional spacetime – a bizarre and marvellous idea that dispensed with the need for either action-at-a-distance or a 'mechanical' mechanism of gravitational transmission. The famous analogy of a bowling ball on a trampoline illustrates the idea: the heavy ball curves or warps the material around it, whereas a tennis ball barely makes a dent at all. But if the tennis ball is placed near the bowling ball, it will slide downhill towards the heavier ball – as if attracted by the heavier ball's 'force' of gravity. Substitute spacetime for the trampoline, and you have a sense of why Einstein asserted that gravity is not a traditional force but a curvature in the very fabric of spacetime.

In trying to make sense of the world, however, verbal language can be both a help and a hindrance, as Maxwell pointed out to those of his colleagues who were attached to the idea of mechanical light waves. And yet – as Newton, Einstein and Mawell realised – in a sense theoretical physics is ultimately about language, especially mathematical language, rather than about 'reality' itself. After all, 'reality' is far more complex than we can ever perceive with our physical senses, or comprehend with our everyday sense of intuition. Which means that no theory is the last word on reality. General relativity is still the most accurate theory of gravity in existence, but it is not perfect, especially at the subatomic level. Maxwell's theory, too, needed adapting in order to apply at this level, but so far there is no quantum theory of gravity to rival the phenomenonally successful theory of quantum electrodynamics (or QED). Consequently, some physicists are working on alternative theories of gravity – and others are still looking for the ether, and are questioning Einstein's conclusions about the speed of light. Whatever happens with these particular lines of enquiry, it is expected that one day general relativity will be incorporated into a larger theory, just as Newton's theory was incorporated into Einstein's. Such twists and turns and changes of fortune always accompany the search for meaning, and for an ever-deeper understanding of our world, and they will continue, of course, for as long as human curiosity endures.

# APPENDIX

## CHAPTER 4: Émilie and Voltaire's Academy of Free Thought

### Figure 1: Kepler's second law

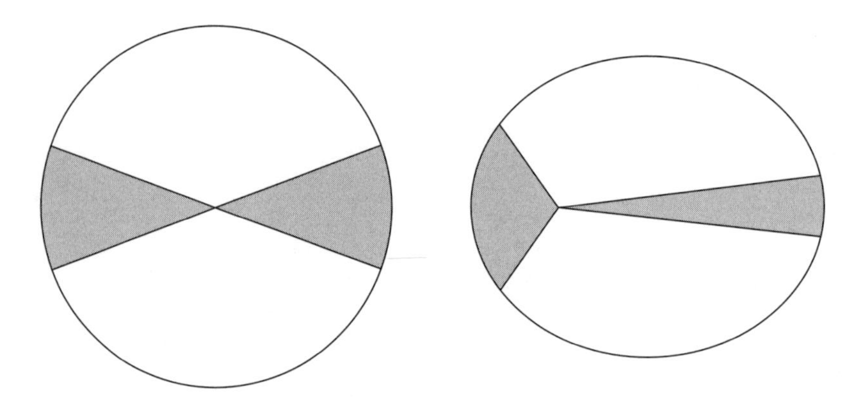

Consider the two shaded sectors in each diagram: those shown in the circle originate from the centre and have the same area and the same arc length, but those in the ellipse originate not from the centre but from the 'focus' of the ellipse (where the sun is located), and so two sectors with the same area will have different arc lengths. (The diagram is illustrative only, not accurately drawn to scale.) In order for planets to make equal areas in equal times, they must travel faster when they are closer to the sun than when they are further away – because the arc length of the sector near the sun is greater, so the planet must travel a longer distance in the same time.

## *Elements of Newton's philosophy*: Kepler's third law

In the narrative, I spoke of the 'radius' of a planetary orbit, but, more technically, I mean the length of the semi-major axis ($r$) of its orbit – that is, the radius of the orbit in its longest, most flattened direction. The 'period' of an orbit is $T$, the time taken for one revolution about the parent body. The illustrative example given in the *Elements* was this:

'Jupiter has four satellites, the nearest being 2⅚ Jupiter diameters [away from the centre of Jupiter], and it orbits Jupiter in 42 hours; the last moon revolves around Jupiter in 402 hours. I want to know how far this last satellite is from the centre of Jupiter. To do this, I use the following rule: as the square of 42 hours, the period of revolution of the first satellite, is to the square of 402 hours, the period of the last, so is the cube of the two [orbital radii] . . .'

The equation embedded in this passage is $\frac{42^2}{402^2} = \frac{(2\frac{5}{6})^3}{r^3}$, where $r$ is the distance of the fourth moon. Now, $\frac{42^2}{402^2}$ is approximately 0.0109, so $r$ can be found by rearranging the equation and taking cube roots. The answer Voltaire gave for the distance to Jupiter's fourth moon was '12-and-⅔ Jupiter diameters', which was a good estimate for someone who had to estimate cubed roots without a calculator: the solution to the above equation is closer to 12.77. (Note: in addition to its four large moons, Jupiter has many smaller natural satellites.)

To show how the usual form of Kepler's law follows from this example, the above equation can be written symbolically as $\frac{T_1^2}{T_2^2} = \frac{r_1^3}{r_2^3}$, which can be rearranged to give $\frac{T_1^2}{r_1^3} = \frac{T_2^2}{r_2^3}$. More generally, Kepler's third law can be written as $\frac{T^2}{r^3} = a$, where $a$ is constant for every planet. Newton's law of gravity shows that this constant depends on the mass of the sun and on the gravitational constant $G$. Similarly, each family of moons around a planet has its own value for $a$, which depends on the mass of the planet and on $G$. (See Resnick and Halliday, p 402, for the derivation.) This means that if the distance and orbital period of one of Jupiter's moons is known, then Kepler's law can be solved for $a$, and so the mass of Jupiter can be found. Similarly, the mass of the sun can be calculated if the data for one of its planets is known.

Also, knowing $T_1$ and $r_1$ – say, the year and orbital radius for the earth – and knowing the year, $T_2$, for any other planet, such as Venus, $\frac{T_1^2}{r_1^3} = \frac{T_2^2}{r_2^3}$ can be solved as before, to find the distance, $r_2$, from Venus to the sun.

## CHAPTER 5: Testing Newton: the 'New Argonauts'

### Measuring the earth: why the north–south distance between two lines of latitude a degree apart is greater at higher latitudes

This is because latitude and longitude are ancient concepts, so angles of latitude are measured as if the earth were spherical: the equator is defined to be at zero latitude and the North Pole is ninety degrees north (the South Pole is at ninety degrees south). All other points on the surface of a spherical globe are measured from the centre of the earth, by measuring the angle between a radial line to the equator and the radial line to the designated point on the surface. If the earth were flattened near the poles, however, then suppose A is a point on the flattened surface, somewhere near the Arctic; a radial line through A would have to be extended out a little further to reach the corresponding point (call it B) on the imaginary spherical surface, because the earth is 'squashed' inwards. This means A would actually be at a lower latitude than B, and so you would have to travel further north along the earth to reach the designated latitude at B.

The technique of measuring latitude from the stars with reference to a 'celestial sphere' centred on a spherical earth was handed down from the ancient Greeks, via the Alexandrian astronomer Ptolemy; however, it is entirely natural to see the sky as a huge spherical dome, and for millennia ancient mariners used a version of this technique without ever having heard of Ptolemy. But the value of having accurate, universally accepted measurements of latitude and longitude is obvious, and one of La Condamine's tasks was to try to determine the exact location of the equator, the reference line of zero latitude.

## The period of a pendulum

The period of a simple pendulum is $T = 2\pi\sqrt{\frac{l}{g}}$. The oscillations of a physical pendulum with small amplitude can be analysed in terms of an idealised simple pendulum. For further analysis, see Resnick and Halliday, pp 360 ff.

## Trigonometry and triangulation

Triangulation makes use of basic properties of triangles, including the fact that the sum of the three interior angles of a triangle is always 180 degrees, so that if you can measure any two angles, you automatically know the third. It also uses the trigonometric relationships between angles and sides in a triangle. To take a simple example, recall from school that if you have a right-angled triangle with a second angle $\theta$ and a hypotenuse of length $h$, then the other two sides can be found from the sine and cosine of $\theta$: the opposite side equals $h\sin\theta$, and the adjacent side equals $h\cos\theta$. This is also a way of *defining* sine and cosine, but sines and cosines for various angles are already tabulated, so in this example, if you know the length of one side, $h$, and you know two of the angles, $\theta$ and the right angle, then you can calculate the length of the other two sides of the triangle.

An important trigonometric relation for when the triangle is not right-angled is the cosine rule: in any triangle with sides of length $a,b,c$, if the angle opposite the side of length $c$ is $\phi$, and if this angle is known, and the lengths *a* and *b* are known, then the length *c* can be found from the 'cosine' rule: $c^2 = a^2 + b^2 - 2ab\cos\phi$. In a right-angled triangle, $\phi$ is 90 degrees, and $\cos\phi$ is therefore zero, so the cosine rule gives Pythagoras's theorem: $c^2 = a^2 + b^2$. In geophysical triangulation, the triangles are 'three-dimensional', but they can be analysed in terms of their component planar triangles.

**CHAPTER 7: The nature of light: Émilie takes on Newton**

**Figure 2: The law of reflection**

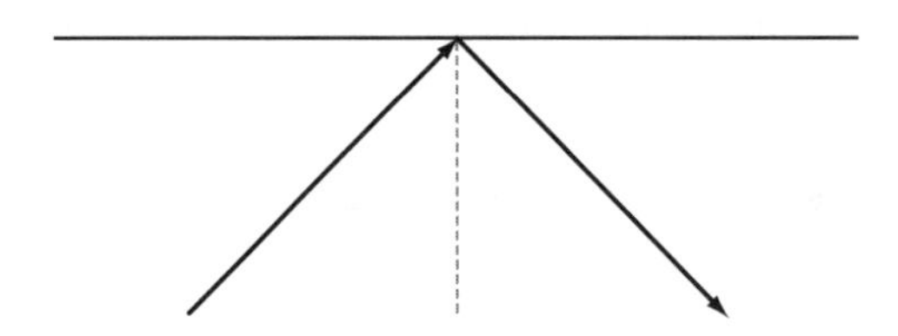

The reflected light ray leaves the horizontal surface with an angle of the same magnitude as it had when it arrived at the surface.

**Figure 3: A spherical mirror focuses light by reflection**

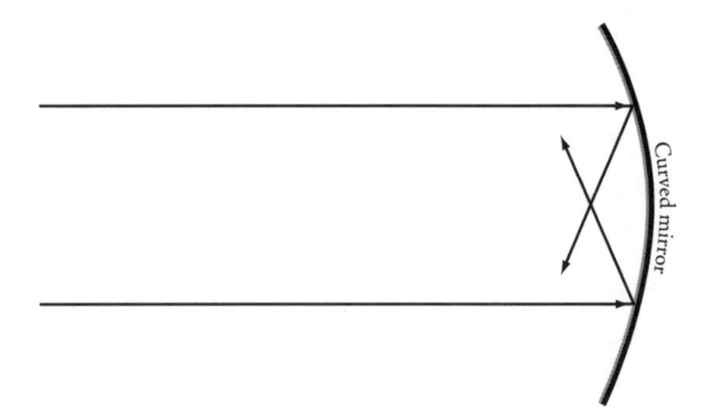


Incoming light rays are reflected so they all meet at a focal point; only two rays are sketched here (and the angles are not exact), but in practice the angle of the incident light and the curve of the mirror are such that all the rays are focused at the same point.

**Figure 4: Refraction of light**

The following diagram sketches the situation for a single reflected light ray travelling from the bottom of a pencil or stick dipped in water out into the air and on to an observer's eye.

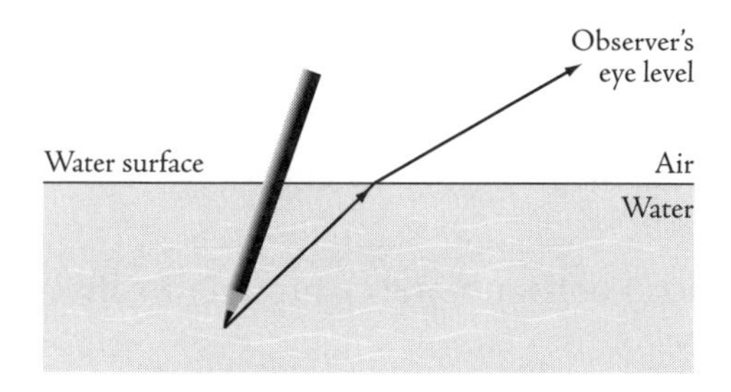


**Figure 5: The light ray is bent on refraction but we 'see' the pencil as bent**

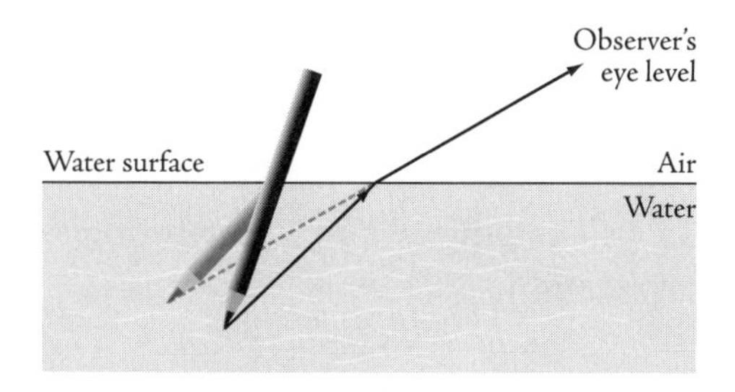


Light travels at different speeds in air and water. But we are so used to seeing light travelling in straight lines that our eyes interpret the refracted light ray to be straight, so we 'see' the *object*, not the light ray, as being bent.

**Figure 6: Refraction enables a lens to focus light**

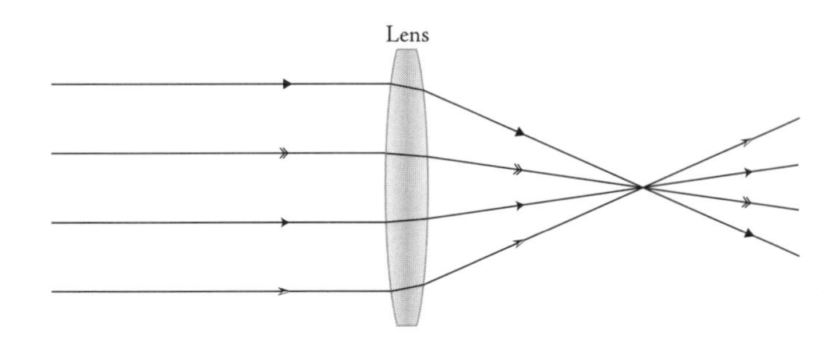


The incoming light rays are bent by the lens so that the rays meet at the focal point. The diagram is schematic only.

**Figure 7: Voltaire's ball-and-water model of refraction**

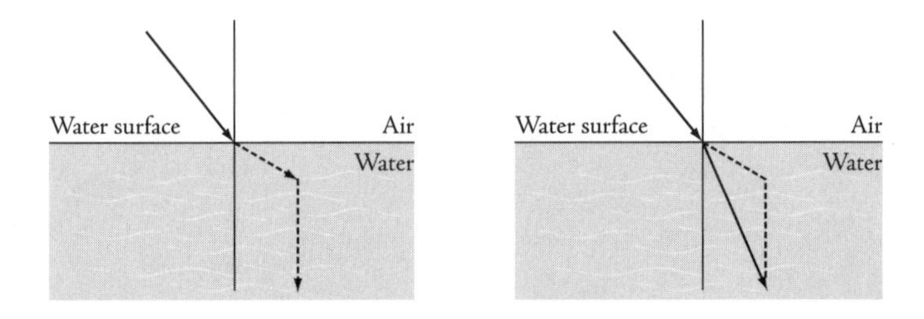


Throwing a heavy ball obliquely into water produces an 'effective' path that mimics refraction.

## CHAPTER 9: Mathematics and free will

### Calculus: Problems with the infinitesimal

In the 'differential' branch of calculus, a 'derivative' is the name given to a 'rate of change' or gradient. Derivatives are the basis of the study of motion, because speed is the rate of change of distance with respect to time, so the instantaneous speed of a moving object is the 'limit' of the ratio of the distance travelled per infinitesimal interval of time, as measured at the required 'instant'. Similarly, the acceleration of an object at any instant is the limit of the ratio of the change in speed over an infinitesimal amount of time. Such ratios are often very close to 0/0, but the question is, how close? One of the criticisms aimed at Newton and Leibniz was that there did not seem to be a transparent way of quantifying the idea of 'infinitesimal'.

Finding derivatives is a more sophisticated process than the term 'ratio' suggests, but the problem with infinitesimals is readily illustrated by the problem of 0/0. It could be argued that because 3/3 = 1, 2/2 = 1, and 1/1 = 1, then 0/0 = 1, too. However, when you divide an ordinary number like 1 or 2 by a very small number, you get a very large number: for example, 1/0.1 (or 1 ÷ 0.1) is equal to 10, but 1/0.000001 is equal to a million. As you divide 1 by smaller and smaller numbers, you get larger and larger numbers, so it seems that 1/0 would be an *infinitely* large number. Consequently, division by zero itself is not allowed in mathematics, but clearly, dividing by an 'almost zero' number is problematic if you don't know how close to zero the divisor should be. And dividing one 'almost zero' number by another leads to even more confusion: 0.001/0.001 = 1, whereas 0.001/0.000001 = 10000, while 0.00001/0.001 = 0.01; so, does the 'limit' – as the ratio approaches 0/0 – equal 1, 0, or some other number, or does it approach infinity (∞)?

A sophisticated theory of 'limits' is needed to work out what happens to ratios (and to functions in general) when the numbers involved tend to a 'limiting' value like zero or infinity. Using this theory, it turns out that the 'indeterminate' ratio 0/0 is, indeed, sometimes ∞ (or 'undefined'), sometimes 0, sometimes 1, and sometimes it is another finite number – depending on the context of the problem. The theory enables the right answer to be found in the given circumstances. Limit theory also gives a precise definition of the underlying concept of a derivative, something neither Newton nor Leibniz had provided adequately.

## The benefits of Leibniz's notation

Leibniz called these infinitesimal quantities 'differentials', and he denoted infinitesimal distance and time intervals as $dx$, $dt$, and so on. Then, in the familiar notation of today's school calculus classes, he represented 'derivatives' as $\frac{dx}{dt}$, and so on, where the $d$ denotes a tiny increment, so that $\frac{dx}{dt}$ denotes a tiny increment of distance ($x$) divided by an increment of time ($t$), so it represents an instantaneous speed.

The problem with derivatives, however, is not only the problem of limiting ratios, but also the fact that derivatives are conceptually far more complex than the simple algebraic ratios I used above. Instantaneous rates of change are not really fractions, $dx \div dt$; rather $\frac{d}{dt}$ is an *operator* that is applied to a *function* of time, called $x$, not to a *number* $x$. But the beauty of Leibniz's notation is that in many situations it allows derivatives to be treated *as if they were ordinary fractions*. For example, the 'chain rule' gives the rule for 'changing variables' in a derivative, and it can be expressed as if the two $dx$ terms actually cancel as in fractions: $\frac{dy}{dt} = \frac{dy}{dx}\frac{dx}{dt}$. Conceptually, the right-hand side of the equation is really a *single* derivative of a *composite* function, $\frac{d}{dt}(y(x(t)))$, just as on the left-hand side, but to calculate this composite derivative it is often easier to use the version on the right, with its algorithmic device of a 'fraction-like' cancelling in a product of two simpler derivatives. A practical example of the utility of Leibniz's 'ratio' notation in the chain rule is given below in the section on deriving *vis viva* from Newton's force law.

By contrast, Newton spoke not of differentials and derivatives but of 'moments' and 'fluxions': his 'fluxions' were represented not as ratios but as letters with a dot on top: for instance, instead of Leibniz's $\frac{dx}{dt}$, Newton wrote $\dot{x}$. (In earlier works he used $p$ or other letters: lower case for derivatives, upper case for antiderivatives.) His 'moments' refer to 'moments of time', and they were represented not with the help of Leibniz's differential *d* but by the Greek letter *o*, which looked rather like a zero but was not zero – just as infinitesimal 'moments' were not quite zero in magnitude. As a simple illustration of the intuitive value of Leibniz's notation, consider the relationship between speed, time and distance. Speed (which is the magnitude of velocity) is a change in distance per time, and it can be written as $v = \frac{dx}{dt}$. If you want to find the distance travelled at this speed in an instant of time, you can write $vdt = \frac{dx}{dt}\,dt = dx$. In Newton's notation, $\frac{dx}{dt}\,dt$ would be $\dot{x}o$, which does not so readily suggest that this quantity is simply an increment of distance.

However, in physics today, Newton's compact notation $\dot{x}$ is often used specifically for velocities (and his $\ddot{x}$ is used for 'second derivatives' such as acceleration), although, in general, Leibniz's utilitarian notation reigns supreme.

## A brief comment on calculus algorithms

The standard approach to teaching 'integral calculus' at school begins by finding the area under a curve drawn on a page ruled with the usual $x$ and $y$ axes. On a given domain, the area between the two axes and the curve is found by approximating the area by a series of tiny vertical rectangles whose top right-hand (or left-hand) corner touches the curve. When these rectangles are considered to have an *infinitesimal* width, the sum of the areas of all the rectangles gives the *exact* area under the curve. The algebraic part of integral calculus comes into play because the geometric construction of rectangles yields an algorithm that dramatically simplifies the calculation of areas of complicated shapes. Using the simple algebraic formula for the area of a rectangle – that is, area = height multiplied by width – and using the algebraic formula for the curve in question, the algorithm giving the total area under the curve can be found.

For example, if the curve is a parabola with equation $y = x^2$, the area of a tiny rectangle is $ydx$ (height by width) or, in this case, $x^2dx$. Summing an infinite number of these tiny areas requires the evaluation of an infinite series, but it turns out to have a remarkably simple answer. In fact, for the area under any curve whose equation is similar to a parabola in that it is a power of $x$, the general algorithm is $\int x^n dx = \frac{x^{n+1}}{n+1}$, where the integral sign $\int$ represents an 'S' for 'infinite sum', in Leibniz's notation, and where, technically speaking, an arbitrary constant must be added to the right-hand side if the domain of integration is not given. The area bounded by the parabola, for which $n = 2$, is thus $\frac{x^3}{3}$ (+ C, the arbitrary constant). If you want the specific area between values of x that are, say, 0 and 2, then C is not needed and the area is $\frac{2^3}{3} - \frac{0}{3} = \frac{8}{3}$. A practical application of this algorithm is given below, in finding the formula for kinetic energy or *vis viva* from Newton's force law. (Below, I use powers of $v$ not $x$, but the rule is the same.)

There are other algorithms for different types of curves. The beauty of it is that these algorithms can be used to calculate any infinite sum of infinitesimally small quantities, not just areas.

There are algorithms for differential calculus, too: for example, for a function like a parabola or other power of $x$, the algorithm is $\frac{dy}{dx} = \frac{dx^n}{dx} = nx^{n-1}$. (For the mathematical derivation of calculus algorithms, see any calculus textbook, for example, Thomas, pp 59–61, 66–70.) Note from the algorithms that differentiation and integration are 'inverse' operations: the one 'undoes' the other. Consequently, integration is sometimes called 'anti differentation'.

These algorithms, devised independently by Newton and Leibniz (and anticipated even earlier), are extraordinarily simple, but, as I mentioned, it took another two centuries for mathematicians to rigorously understand the conceptual basis of calculus.

### Both *vis viva* and momentum follow from Newton's definition of force

Newton's second law of motion, or his general law of force, is that force is proportional to the (instantaneous) change of momentum (or quantity of motion). Mathematically, this can be written as $F = \frac{d}{dt}(mv)$. On integrating this equation with respect to $t$ (that is, 'antidifferentiating' or 'undoing' the derivative in the expression for force), you obtain the momentum, $mv$. In other words, assuming mass is constant, and including the arbitrary constant of integration (see Resnick and Halliday p 212 for the physical role of this constant), and using the integration algorithm for the 'power function' $1 = v^0$:

$$\int Fdt = \int \frac{d(mv)}{dt}\,dt = m \int \frac{dv}{dt}\,dt = m \int dv = mv + C.$$

Leibniz stated this result in words in his 1695 *Essay on Dynamics*, extracted in Wiener (ed.), p 124.

On the other hand, integrating Newton's law with respect to distance ($x$) rather than time yields the formula for *vis viva* rather than momentum: the following line of calculations uses the definition of velocity as $v = \frac{dx}{dt}$, and it also uses both the chain rule and the algorithm for integrating a power function (in this case $v = v^1$). It also assumes mass is constant, which is valid in most everyday applications:

$$\int Fdx = \int \frac{d(mv)}{dt}\,dx = m \int \frac{dv}{dt}\,dx = m \int \frac{dv}{dx}\frac{dx}{dt}\,dx = m \int \frac{dv}{dx}\,vdx =$$

$$m \int v\frac{dv}{dx}\,dx = m \int vdv = m(\frac{1}{2}v^2) + C.$$

## CHAPTER 10: The re-emergence of Madame Newton du Châtelet

### Creating the theory of gravity (a modernized account)

At the heart of Newton's argument is a comparison of the motion of the moon around the earth with that of a falling body here on the surface of the earth. Firstly, consider the moon. Newton derived a simple formula expressing the acceleration of a body that is moving in a circle of radius $r$ with constant speed $v$, namely $a = \frac{v^2}{r}$, which he used to find the acceleration of the moon as it revolves around the earth. (Huygens independently derived this formula in a different context.) The radius of the circle in this case is the average distance of the moon from the earth, which astronomers had already measured approximately. The moon's speed $v$ could then be calculated simply by dividing the distance it travels in one circular orbit by the time taken – that is, dividing $2\pi r$, the circumference of the orbital circle, by 27.3 days, which is the average time it takes the moon to orbit the earth once. Knowing $v$ and $r$, Newton was then able to use his formula to calculate the moon's approximate acceleration.

In the next part of his argument, Newton compared the moon's acceleration with the acceleration of a falling body here on earth. To do this, he used Galileo's famous discovery that all bodies near the earth's surface fall with virtually the same constant acceleration. Physicists today use the letter $g$ to denote the acceleration during gravitationally induced freefall: in modern units (to one decimal place), $g$ = 9.8 metres per second per second near the surface of the earth. (The slight variations in gravity at different latitudes and different earthly altitudes – up to about 16,000 metres – only affect $g$ when it is calculated to more than one decimal place; see Resnick and Halliday p 395, tables 1 and 2.) Newton's calculation showed that the moon's acceleration was about $\frac{g}{3600}$.

Now, it happens that the moon is about 60 times as far from the centre of the earth as we are on the surface of the earth; in other words, the distance between the centre of the earth and the moon is 60 units, where 1 unit is the radius of the earth. So, the square of the distance between the earth and moon is 3600 times the square of the distance from the earth's centre to its surface, which means that if the earth's gravity reaches as far as the moon, and if the inverse-square law holds, then $\frac{g}{3600}$ is exactly what you would expect for the amount of *gravitational* acceleration of the moon towards the earth.

With these two calculations, Newton had shown that the *actual* acceleration of the moon was $\frac{g}{3600}$, and so was its *theoretical gravitational* acceleration, which meant that the earth's gravity was entirely sufficient to be the sole cause of the moon's orbital motion. (Inaccurate data for the earth's radius meant his calculations were not quite so perfectly matched as this, and it has been suggested he slightly fudged his results (see Cohen [1999], for example, pp 369–70); but, as Curtis Wilson pointed out (ibid., p 205, footnote 17), 'a good enough agreement [between theory and experiment] would have been obtained without the fudging'. Today, we know that the development of a theory requires intuition as well as logic, and that theory and experiment can never give exact agreement, but, as Cohen pointed out (ibid., p 369), what Newton was doing in constructing the very first quantitative theory in history was so novel that he had 'no guides to follow, no standard procedures'. No wonder he felt his theory would only be accepted if it was in perfect agreement with the physical facts.

But how did Newton arrive at the inverse-square law in the first place? As I mentioned in Chapter 2, he proved mathematically that the force producing *any* elliptical motion *must* be an inverse-square force (if the source is at the focus of the ellipse). He also proved this for motion along the other eccentric 'conics', parabolae and hyperbolae, which are the paths taken by many comets. The proof for the approximately circular motion of the various moons is the easiest to follow; it uses the formula for circular acceleration together with Kepler's third law, as follows (although Émilie pointed out that this proof only applies to systems of moons, not to single moons like the earth's, because Kepler's law is relative; hence Newton needed to demonstrate the inverse-square law for our moon by a different argument, see the *Principia*, Book 3, Propostition 3, cf Propostion 1, although the following derivation is more transparent than the proof of Proposition 1).

**Deriving the inverse-square law for celestial circular motion**

Units of time are those of distance/speed, so the orbital period $T$ is proportional to $\frac{r}{v}$. (In other words, for circular motion, $v = \frac{2\pi r}{T}$, so that $T$ is proportional to $\frac{r}{v}$.) Also, Kepler's third law says $T^2 \propto r^3$, where $\propto$ denotes proportionality.

With the above expressions for $T$ as 'givens', now insert the formula for the acceleration of circular motion at constant speed – that is, $a = \frac{v^2}{r}$ – into Newton's second law of motion, $F = ma$, where $m$ is the mass of the object (say, the moon) that is moving under the applied force $F$ (for constant mass, the formula 'mass times acceleration' is equivalent to the change in momentum referred to earlier in connection with Newton's definition of force). You will obtain the following:

$F = ma = \frac{mv^2}{r}$. Multiply numerator and denominator by $r$ to rewrite this as follows:

$F = ma = \frac{mrv^2}{r^2}$. From the first expression for $T$ given above, $\frac{v^2}{r^2}$ is proportional to $\frac{1}{T^2}$, so we can rewrite the equation again:

$F = ma = \frac{mrv^2}{r^2} \propto \frac{mr}{T^2}$. But from Kepler's third law, $T^2 \propto r^3$, so we can rewrite again:

$F = ma = \frac{mrv^2}{r^2} \propto \frac{mr}{T^2} \propto \frac{mr}{r^3}$. Cancelling the factor of $r$ in numerator and denominator, we end up with the inverse-square law,

$F \propto \frac{m}{r^2}$.

The proportionality factor in Newton's inverse-square law is $GM$, so the exact law is $F = \frac{GMm}{r^2}$ where $G$ is a universal constant (first measured by Cavendish) and $M$ is the mass of the other object (such as the earth) whose gravity is the source of the force causing the motion of the accelerating body (the moon in this case).

($G$ is constant in Newtonian and Einsteinian gravitational theory, but see Misner, Thorne and Wheeler, pp 1121–5, for $G$ in other theories. Note, too, that Einstein's is still the simplest and best theory we have so far, despite the fact that it does not explain quantum effects or singularities such as the centre of black holes or the Big Bang itself.)

## A sketch of the proof of the 'superb' theorem, Proposition 70, Book 1 of the *Principia*

Recall this theorem says that for a particle placed anywhere inside a *hollow* sphere (that is, inside a spherical shell), there is *no* net gravitational pull from the shell. It is easiest to see this through Newton's geometrical way of thinking, rather than calculus algorithms. Visualise the shell in the usual two-dimensional way as a circle on the page with a 'north pole' at the top. Then, imagine the interior particle is placed at the centre of the sphere, so it is equidistant from every point on the spherical shell. A tiny piece of the shell at the 'north pole' will attract the central particle 'up' towards itself, while an equally small piece directly opposite, at the 'south pole', will pull the central particle 'down'.

In other words, these two forces are equally strong, because the distance to the centre is the same in each case, but they act in opposite directions, so they cancel each other out and the central particle feels no force at all. All the matter on the surface can be divided up into infinitesimally small pieces and paired off on opposite sides in this way, so there is no net force of gravity from the shell on the central particle. Newton also showed that this kind of pairing and cancelling works when the internal particle is placed at any point inside the shell, not just the centre.

When the sphere is full of matter, however, there are so many tiny pieces of matter to consider that this kind of cancelling does not occur; instead, as I mentioned, Newton found that the average of all the individual inverse-square forces between all the pairs of particles produces, rather miraculously, a simple proportionality law.

## CHAPTER 14: The long road to fame

### Modern definitions of the second, the metre and the inch

The 24-hour time interval between one high noon and the next – that is, the period of the earth's rotation on its axis – is still the standard measure of everyday time. The second is defined to be $\frac{1}{60} \times \frac{1}{60} \times \frac{1}{24}$ of this 24-hour interval. But since 1967, a more accurate periodic phenomenon has been used to define the second more precisely for scientific purposes: the frequency of radiation emitted from an atom of cesium 133.

Although the original metre bar is still accurate for most purposes, in 1983, the international General Conference on Weights and Measures redefined the metre more precisely as the distance travelled by light through a vacuum in $\frac{1}{299{,}792{,}458}$ of a second. This choice reflects the fact that the speed of light, denoted by the letter $c$, is a universal constant, the same for all observers anywhere in the universe. So in metric units, $c$ is defined to be exactly 299,792,458 metres per second (in a vacuum). This definition of the metre is not only more accurate than a physical measurement along a meridian of the earth, it is also able to be calibrated at individual laboratories, rather than requiring the existence of a unique measuring bar that is vulnerable to damage or destruction.

As for ancient measures like miles and feet, the mile was originally derived from the measure of one thousand paces of a Roman soldier. Today, these ancient British units, which are still used in the US, are defined with respect to the standard metric system: one inch is defined to be exactly 2.54 centimetres, and the other correspondences carry through: twelve inches in a foot, three feet in a yard, 5,280 feet in a mile, and so on.

(Sources and/or further reading: Taylor and Wheeler, p 58; Resnick and Halliday, p 5.)

## CHAPTER 15: *Mechanism of the Heavens*

### Using calculus to minimise physical quantities

The method is essentially the same whether you are finding a minimum or a maximum, and it depends on the idea of a gradient, or slope. A gradient is a rate of change (or ratio), and rates of change are the 'derivatives' referred to in differential calculus. For a straight line, this gradient is just the ratio of the change in height to a given horizontal change. For a curved line like a cycloid – or a parabola or semicircle or any other curve – the gradient at any point is defined to be the slope of the straight line that is *tangent* to the curve at that point; the tangent can be drawn by laying a ruler on the curve at the required point, and angling the ruler so it lies as close as possible to the curve but does not touch any other point on the curve.

To see how this helps to find a minimum or maximum value of the curve, imagine a curve that is shaped like an arch, or a rainbow. In your mind, visualise the point at the very top of the arch, and then draw a horizontal line through it. This is the tangent line through the highest (or maximum) point of the curve. The very top point of any curve that is shaped like an arch is the *only* point where the tangent will be horizontal. This means that identifying the existence of a horizontal tangent is mathematically the same as locating the maximum of the curve – or the minimum, because if you imagine turning the arch upside down, its turning point is at the bottom of the curve, at its lowest or minimum point. Being horizontal means this tangent has no slope at all, so its gradient or derivative is zero.

Not all curves have such a convenient shape for visualising maxima and minima (for instance, not all curves have a single turning point, but may have several 'local maxima' or 'local minima'), but the point I am making is that by analogy with the arch, *any* mathematical function – not just those describing a parabolic or semicircular arch – can be maximised or minimised by setting the algebraic expression for its derivative equal to zero. Often the context makes it clear whether you have a maximum or minimum, but there is a mathematical way of deciding which is which.

Calculus of variations incorporates this property of ordinary calculus in such a way that, out of all the possible curves a body could trace, the unique curve can be found that minimises some aspect of the body's motion. In the brachistochrone problem, mathematicians need to minimise $t$, the time taken when the body slides down an arbitrary chute from point A to point B. When the mathematical expression for $t$ is minimised in this way, a new equation is generated, which gives the physical profile of the unique chute that will allow the body to slide down in the shortest possible time – namely, the cycloid. (In calculus of variations, the minimum is defined by a zero 'variation' rather than a zero derivative, but this can be 'converted' into a differential condition [via the Euler-Lagrange equations, for example]. Mary and Laplace explained the conditions where it was acceptable to interchange the terms 'variation' and 'derivative'.)

## A simple example of how the principle of least action produces Newton's laws of motion

Following Laplace, Mary gave a very simple example that illustrates the power of the principle of least action to produce the relevant Newtonian laws or 'equations of motion' – in this case, the 'law of inertia', which says that when a body is moving with a constant velocity, without any accelerating force, it moves in a straight line, and it continues to move at the same speed in the same straight line until some force stops or diverts it. For instance, if you give a billiard ball a direct hit with a cue, the impact causes the ball to move in a straight line with a constant speed (assuming the billiard table is so smooth as to have no frictional force, and assuming it doesn't hit anything to give it another impulse or force). Mary showed that the same result can be derived purely mathematically from the principle of least action – without reference to observation or to physical concepts like inertia. Using velocity multiplied by distance for the action, the argument goes like this: if the velocity is assumed to be constant, then it cannot be changed or 'minimised', and the principle of least action reduces to minimising just the distance; but the shortest distance between two given points is a straight line, and so the principle of least action says that when no forces are acting on it, a body moving at constant speed will move in a straight line.

In other words, the law of inertia follows as a *mathematical consequence* of the principle of least action, rather than from physical observations – and the same applies to more complicated equations of motion. Since these equations determine the trajectory of the object, this result suggested the actual path taken by an object moving under a force is the one that minimises or 'economises on' the 'action'.

Note that Mary and Laplace used the action *integral*, $\int v ds$, which is essentially the same as velocity times distance ($s$) if the velocity is constant.

## Both momentum and energy are conserved: an illustration

In a system of particles free of external forces, the *total* momentum of all the particles before and after an interaction is the same. For instance, in the case of falling motion – which concerned Leibniz, see Wiener (ed.), p 136 – the 'system' includes the falling object and the earth: Newton's law of gravity operates mutually, so the falling body gravitationally attracts the earth upwards just as the earth pulls down on the body. But force is proportional to the change in momentum, so the falling body's momentum is equal and opposite to the earth's at any point during the motion. (The earth is trillions of times more massive than an everyday falling object like a ball, so if its momentum is equal in magnitude to that of the ball, the earth's velocity towards the ball is imperceptible, because it must be trillions of times smaller than the ball's falling speed.) This equal and opposite momentum of the earth and the ball means the total momentum of the system is unchanged: it was zero before the ball was dropped and it remains zero throughout the fall.

The mechanical energy of such a system is also conserved: the one does not preclude the other, but Leibniz was right in that momentum and energy are different quantities, and care needs to be taken in interpreting them. In analysing falling motion in terms of energy, in general only the potential and kinetic energies of the falling body are considered: although the earth itself moves in response to a falling ball's gravitational force, its speed is so tiny that when this speed is squared, the earth's kinetic energy is negligible. (See Resnick and Halliday, pp 197–8, for the calculations.)

Note, however, that 'mechanical' energy is only conserved when motion is due to 'conservative' forces like gravity, in the absence of other forces like air resistance or friction (although conservation of *total* energy takes account of *all* forces); in practice, air resistance is often negligble. Note, too, that momentum is not conserved if the system is not 'isolated' from significant extraneous forces: for instance, in collisions, friction can generally be ignored because it is minute compared with the force of the collision – but if a light-weight ball were pushed towards a stationary ball on a table covered with rough material, friction might be so great that the ball stopped before it even hit the stationary ball. Clearly momentum is not conserved in this case. (Sources and/or further reading: Resnick and Halliday, pp 155, 197–8, 214, 222–3.)

## Why Leibniz and Émilie thought Newton did not accept *vis viva*; Newton's 'Query 31'

Émilie had followed Leibniz's misguided attack on Newton, writing in her *Fundamentals of Physics*, 'The very name of Newton is essentially an objection against living force.' This was because 'Query 31' of *Opticks* was published more than two decades after the *Principia*, and a decade after Leibniz's work on *vis viva*, and yet Newton had not mentioned 'living force' or its equivalent. To add to the confusion, he *had* spoken of the fact that momentum is not conserved in the universe, so that the amount of 'motion' in the world was 'always upon the decay', as he put it.

This was sloppy terminology – even in the *Principia* Newton frequently wrote 'motion' when he meant 'quantity of motion' or momentum, so it was not entirely clear whether he meant that momentum or motion itself was 'on the decay'. Either way, Émilie – following Leibniz – conflated this with Newton's later statement in the 'Query', that the solar system would want re-forming (or 'winding up' in Leibniz's sarcastic words). Which is surprising, because Newton had concluded his 'Query' by saying: 'Seeing the variety of motion which we find in the world is always decreasing, there is a necessity of conserving and recruiting it by active principles, such as are the cause of gravity . . . and fermentation.' He mentioned fermentation to 'explain' how living organisms maintained their ability to move and to keep themselves warm. He continued, 'And if it were not for these principles, the bodies of the Earth, planets, comets, Sun, and all things in them, would grow cold and freeze . . . and life would cease, and the planets and comets would not remain in their orbs.'

This was an intuitive groping towards the idea that there is something else besides the Cartesians' all-encompassing momentum that is needed to keep the universe going, to conserve it, and that this 'something' is related to both heat energy and sources of motion. In fact, Newton's references to vegetative decay and transformation – and other such apparently alchemically inspired comments in his 'Queries' – show that his conception of nature was not one of 'dead' forces and mechanistic materialism, but a world full of energy and life force. Nevertheless, he did not appear to have as clear an idea of the concept of kinetic energy as Leibniz did with his 'living force', nor did he so clearly articulate an energy conservation law (although I say 'clearly' in purely relative terms!)

## Conservation of energy theorems in the *Principia*

Newton used the equivalent of his 'force times distance' formula to show – in Propositions 39–41 of Book 1 – that if a body acquires a speed $v$ after falling a certain distance, then the 'work' done by gravity is equal to the kinetic energy gained during the fall. He did not use this verbal language, of course, but his mathematical result is the equivalent of Leibniz's principle of 'living force'. (Actually, Newton's theorems applied to motion under *any* centripetal force, and Proposition 41 was not restricted to vertical motion, so these theorems provide a more general statement of conservation of mechanical energy than Leibniz's principle.) All this was expressed purely mathematically, not in terms of the above transparent modern terminology – Newton did not even use verbal terms as rough as Leibniz's 'living force'. But Newton's proofs of these propositions are instances of where he used calculus concepts more explicitly than usual. His 'work = force multiplied by distance' formula was implicit as the area under his force–distance graph, representing what we now call the work integral. (Recall that the area under a graph is an example of a calculus 'integral'. Recall, too, that integrating force with respect to distance gives the formula for kinetic energy, or rather, for the *change* in kinetic energy: the constant of integration is the key here. See Resnick and Halliday, p 142.)

Very few of Newton's contemporaries and younger near-contemporaries appreciated what he had done in these Propositions. A notable exception was Jacob Hermann, the Swiss student of Jean Bernouilli's older brother, Jacques: he realised that Jean Bernouilli's later derivation of the formula $mv^2$ was the equivalent, in Leibnizian calculus, of Newton's derivation in Proposition 41. In her commentary on the *Principia*, Émilie, too, discussed Proposition 41 (corollary to her theorem XXX, p 141, Book 2 of her translation and commentary), but she did not explicitly connect it with *vis viva*. The first major discussion of how Newton's work related to the conservation of energy was by William Thomson and Peter Guthrie Tait, who pointed it out in 1867, in their book *Natural Philosophy*. (Sources and/or further reading: Cohen [1999], pp 118–22, and Chandrasekhar, Chapter 9. Jacob Hermann on Newton's Proposition 41: Cohen [1999], p 142. See also Boyer, p 434.)

Nevertheless, it is not clear how much physical significance Newton himself attached to his mathematical theorems. Given his additional practical example of machines lifting heavy objects, and his intuitive comment in 'Query 31' about the need for the universe to conserve heat and motion, it is reasonable to suppose he understood the concept of conservation of energy at least as well as Leibniz did. Thomson, Tait and many others believe this to be the case, arguing that Newton simply lacked the verbal language to express his ideas more concretely. On the other hand, it does happen that mathematical equations can predict something so unexpected it is not noticed by its author. Einstein's $E = mc^2$ is the most famous example: he derived the equation simply in order to predict the kinetic energy of an electron, and when he first published it, he did not remark on its extraordinary implication. Some months later, he rectified the situation! Even so, the physical meaning of $E = mc^2$ remained hypothetical for many years, until the discovery of nuclear fission verified the equation, simultaneously unleashing the spectre of nuclear weapons.

Laplace derived the formula $mv^2$ in a much simpler way than Newton (who used calculus concepts within a geometric framework), by integrating Newton's general (second law of motion) formula for force, the way Bernouilli and others had done nearly a century earlier. He obtained an equation that is equivalent to the modern statement 'kinetic energy plus potential energy is constant', although he did not use this language. (Laplace's mathematics included what we call a 'work integral': see p 99 of Bowditch's translation of *Celestial Mechanics*. Émilie derived this result in her commentary on the *Principia*, section XXXVI, Proposition XXII.)

## Joule's experimental proof of conservation of total energy

Joule used an ingenious set-up where two 'weights' were connected to a rotor by a rope and pulley, and the rotor was connected to a set of paddles in a container of water; the rope around the rotor unravelled as the weights fell under the influence of gravity, thus turning the paddles, which agitated the water and caused it to heat up. (As mentioned earlier, care needs to be taken in distinguishing weight from mass, but in everyday language, we use the term 'weight' to define a heavy object.) The work done by the falling weights was calculated from the formula that Galileo, Leibniz and Newton had pioneered – namely, work = force (or weight) times distance fallen – and the heat gained was measured in terms of the mass of the water and its increase in temperature.

Joule then performed numerous other experiments, where the heat was produced in different ways – by vigorously stirring mercury, for example, or from an electrical current through a wire that was then placed in water, so the heat of the electrical energy was transferred from the wire to the water, where it could be measured as in the earlier paddle experiment. Through these meticulous experiments, he showed that the same amount of heat energy was obtained from a given amount of work, no matter how that work was performed; in this way, he established the equivalence of heat and other forms of energy. Furthermore, Joule's and others' experiments showed that at any point in the experimental process, it was, indeed, the *total* amount of energy that was conserved, not each of the separate types of energy. (Sources and/or further reading: Resnick and Halliday, pp 554 ff.)

## An everyday example of 'resonance'

Resonance is the phenomenon whereby an appropriately timed periodic force dramatically amplifies the 'natural' frequency of a system, that is, the frequency it would have if it were simply perturbed once and let go. Those of us with childhood memories of playground swings will have seen this phenomenon manifested in the fact that when you push your legs periodically in just the right way, the swing rises higher with each push. Similarly, if Jupiter and Saturn were repeatedly near conjunction at the same time, the effect would be dramatically increased until their orbits were significantly changed. (See Resnick and Halliday, p 372, for the swing example and for the mathematics of resonance.)

## CHAPTER 17: Finding light waves: the 'Newtonian revolution' comes of age

### Newton's vibrational theory and experiments

I mentioned that Newton's language was extremely rudimentary as he tried to come to grips with the nature of light. For example, he used the word 'fit' – as in seizure – to describe the peaks of his 'wave', and his 'interval between fits' corresponded to a wavelength. But Newton's waves were not like water ripples or vibrations in an ethereal medium. Rather, he was trying to 'explain' how, in passing through the surface of a refractive medium, part of a light beam is bent by the medium and transmitted through it, and part of it is reflected. He suggested the impetus of the incoming light particles, as they hit a surface like glass, caused two intertwined series of vibrations in the glass, and that the vibrations in each series were staggered in such a way as to allow alternate 'fits' of transmission and reflection.

Newton's language may have been laughably imprecise, but his experiments were meticulously detailed. In particular, he made an extensive study of a different kind of interference pattern – not Young's pinhole experiment where the interference pattern is analogous to that made by water ripples, but the phenomenon known as 'Newton's rings'. These coloured rings are made by interference between the reflections and refractions in thin transparent media like soap bubbles; they had been observed before Newton, but he was the first to rigorously examine and describe them.

In Mary's words, Newton studied the rings that occur when 'a plate of glass is laid upon a lens of almost imperceptible curvature, before an open window; when they are pressed together a black spot will be seen in the point of contact, surrounded by seven rings of vivid colours, all differing from one another'. His detailed measurements included the diameters of various coloured rings, the refrangibility (that is, the amount of refraction or bending) of each different colour of light, and the 'intervals of fits of reflection' of each different colour. He found red light had the greatest 'length of fit' and violet the least. It was this data that Young used to make the first ever calculation of the wavelengths of the different colours of light.

(Sources and/or further reading: Newton's vibrational theory and experiments: *Opticks*, for example, Part II, Proposition XII. For a detailed historical analysis of Newton's experiments, see Shapiro.)

## EPILOGUE: Declaring a point of view

### Statistics on women in the mathematical and other sciences

Note that although it found the average proportion of women researchers was about a quarter, the UNESCO study also showed there are some countries, especially in Eastern Europe, with gender parity among scientific researchers. (Source: UIS Bulletin on Science and Technology Statistics, No 3, November 2006: 'Women in Science: Under-represented and under-measured', at www.uis.unesco.org/template/pdf/S&T/BulletinNo3eN.pdf.)

In the epilogue, I mentioned the 2010 statistics for the School of Mathematical Sciences at Monash University, but back in 1984, there were no female staff members, apart from a handful (five per cent) of tutors, and no female doctoral students, although about one in three Masters candidates were women. In the Canadian study on astronomy referred to in the epilogue, in 1991–1995, only four per cent of professors were women, compared with six per cent in 1996–2000; for the same time periods, the number of female postgraduate students increased from seven to seventeen per cent, and the number of postdoctoral researchers increased from twelve to seventeen per cent. In the US, the percentage of female enrolments in science and engineering degree courses increased from around forty per cent in 1999 to close to forty-four per cent in 2008. Globally, however, women are severely under-represented in senior scientific academic and decision-making bodies: for instance, only fifteen per cent of full professors in Europe are women.

Sources and/or futher reading: for the Canadian statistics, see AAS Committee on the Status of Women: 'Women in Canadian

Astronomy: A Ten Year Survey (1991–2000), published online in 2005 at www.aas.org. For the 2010 figures from Monash University, I am most grateful to Linda Mayer. (Thanks, too, to Professor Kate Smith-Miles, the first female Head at Monash's School of Mathematics.) For the 1984 figures from Monash: Gabrielle Baldwin, 'Women at Monash University', May 1985, Monash University publication. For the US statistics, see the National Science Foundation's website, www.nsf.gov/statistics/.

For the under-representation of women in senior scientific positions: this is from a report, 'Mapping the Maze: Getting more women to the top in research', compiled by the European Commission's Expert Group on Women in Research Decision-Making (WIRDEM), discussed in the article, '"No quality without equality": A new report on women in science', from CORDIS, http://cordis.europa.eu/.

# NOTES AND SOURCES

These notes acknowledge my sources, both primary and secondary; only author's surnames or brief citations are given here – the full references are in the Bibliography. I have also included here occasional additional information in order to clarify or amplify a point in the narrative.

I hesitate to single out important sources, because the researches of the numerous scholars mentioned in the notes below have all helped to confirm or extend my own knowledge; nevertheless, let me especially acknowledge Judith Zinsser (on Émilie), Elizabeth Chambers Patterson and Dorothy McMillan (on Mary), I. Bernard Cohen (on Newton's *Principia* and *Opticks*), and Theodore Besterman (editor of Voltaire's correspondence and some of Émilie's). I have found useful ideas, information or interesting tidbits in these and all the sources listed. I have made these notes as detailed as possible, both in acknowledgment and as a guide for readers who want more information.

I also gratefully acknowledge the Bibliothèque Nationale de France and its helpful librarians, and the Monash University Library and its equally helpful librarians (especially Fiona Russell, Barbara Wojtkowski, Yiye (Amy) Tan and Kim Arndell).

## CHAPTER 1: Madame Newton du Châtelet

In her lifetime, Émilie was often known by the name 'Chastellet', which became 'Châtelet' due to an eighteenth-century modernisation of spelling; Voltaire was the first to begin using Châtelet.

'a temperament of fire': Émilie wrote this of herself, in her *Discours sur le bonheur*, p 7. She was referring specifically to her appetite for fine food, but it applies equally to her appetite for life and love, as the *Discours sur le bonheur* and her personal letters show.

The passion and earlier liaisons of baron de Breteuil: Vaillot, pp 24–7; É combining parents' temperaments, ibid., p 29. Créqui on Breteuil library: ibid., p 29.

Émilie on prejudice, intellectual awakening: her own preface to *The Fable of the Bees*. Mlle de Thil and mathematics: Zinsser, p 60. Voltaire on the young Émilie's knowledge: Badinter, pp 68–9.

Émilie on the death of her infant son: É to Sade, *Lettres de Voltaire et de sa célèbre amie*, undated, p 34; also É to Maupertuis: *Lettres Inédites*, Lettre III. On wet nurses: usually babies were taken away from their family home until they were weaned, but É had her babies' nurses live in: Zinsser.

Voltaire to Cideville: quoted in Anne Soprani's introduction to *Lettres d'Amour au Marquis de Saint-Lambert*; also Vaillot, p 77.

Voltaire's poem to Émilie: *Epître à Uranie* (*Letter to Urania, the Muse of Astronomy*), quoted in Vaillot, p 79, Hamel, p 51, full text available online from Wikisource, Epître 45. Note that Zinsser, p 79, refers to an earlier poem of the same name, in which Voltaire is still pleading with his Uranie to leave aside her study for a while and learn the arts of pleasure from him. Faivre, p 26, and Vaillot, p 58, mention an earlier one still (to a different muse), published in 1732 but written a decade earlier, in which Voltaire attacked Christian dogma (partly motivated because his actress lover's body was refused a Church burial on 'moral' grounds).

Voltaire, 'divine Émilie': *Lettres de Voltaire et de sa célèbre amie*, 29 August 1733. Also V to Sade, quoted in Hamel, p 50; V to Cideville, quoted in Vaillot, p 73.

The importance of discretion and the artifice of romance that prevailed in French society in past centuries can be seen in acutely observed contemporary novels such as Balzac's early nineteenth-century *Le Père Goriot*.

Émilie, 'Despite the pompoms': *Lettres de Voltaire et de sa célèbre amie*, undated, p 57. É actually says 'princesses and pompoms'. 'I love study': É to Maupertuis, 24 October 1738.

## CHAPTER 2: Creating the theory of gravity: the Newtonian controversy

Ellipse from inverse-square law: Newton claimed the converse argument – that centripetal inverse-square forces can only yield orbits that are conics – was so obvious he stated it without proof in the first edition of the *Principia* – a fact that Bernouilli pounced on: Westfall, p 742. See also the discussion in Cohen (1999), pp 133–6, on a minor omission in Newton's subsequently added proof. Note that all my references to Cohen (1999) refer to his 'Guide to Newton's *Principia*', which precedes his translation of the *Principia*.

Newton on comets: *Principia*, Book 3, Propositions 40, 41, 42 and their corollaries. Cohen (1999) points out, p 272, that in the first edition of the *Principia*, Newton speculated that comets might bring into our atmosphere some 'vital spirit' necessary to life on earth, and also material that could be chemically converted into animal, vegetable or mineral matter. Cohen also shows that, privately, Newton made more outlandish speculations relating to how God might use comets to produce the biblically predicted 'cataclysm', but, says Cohen, 'Obviously Newton did not enter such extreme thoughts into the otherwise sober and austere *Principia*.' See Cohen (1999), p 273. See also ibid., pp 56–64, for discussion of Newton's alchemy as part of the foundation of his thought, and the extraordinary fact that, nevertheless, he ultimately chose to separate metaphysics and experimental/mathematical physics, thus forging the secular style of the *Principia* and of theoretical physics itself.

Newton on precession of the equinoxes: for more detailed information, including critiques of Newton's methods, see Cohen (1999), pp 205 ff.

Accuracy of Newton's theory of gravity: Penrose, p 198. (To illustrate: if Newton's theory calculates the distance of a communications satellite orbiting the earth at a distance of 35,800 kilometres, the error would be not much more than three metres.)

Aberration of starlight and rain: *Oxford Reference Encylopedia*, and NASA's website for a more detailed view, including Bradley's analogy not with rain but with a flag, whose apparent position changes according to the speed of the observer.

Descartes, Leibniz, Huygens: Descartes adapting Aristotle's celestial vortices: Leibniz, 'On Aristotle's and Descartes's Theories of Matter' (1671), extracted in Wiener (ed.), p 90. Leibniz to Samuel Clarke, ibid., pp 217–19, and on Newton's theory, ibid., pp 276–9. Leibniz to Newton on his 'astonishing' discovery: *The Correspondence of Isaac Newton*, H. Turnbull (ed.), Cambridge University Press, 1961, Vol 3, p 258, quoted in Maglo, p 149.

Huygens on Newton's theory: in Todhunter, p 35, Article 65.

Bernouilli accepting inverse-square law but not attraction: Terrall, p 49, especially footnote 48.

'The Newtonian style': Cohen (1999), pp 60 ff. Others had edged towards this paradigm before Newton: Kepler, Galileo, Gassendi, Huygens (to name just a few of his more immediate predecessors); medieval pioneers like Roger Bacon; and even ancients like Archimedes. But none of them came close to Newton in scope, precision and mathematical sophistication. Newton on God in the *Principia*: Cohen (1971), p 156; God having no place in natural philosophy: ibid., p 245.

**CHAPTER 3: Learning mathematics and fighting for freedom**

Du Deffand and thwarted ambition: introduction to 1994 edition of Du Deffand's *Lettres à Voltaire*. Also, Du Deffand and Voltaire: Krief, pp 282 ff. Du Deffand on Émilie: Hamel, pp 123–4.

Voltaire on Molière: *Epitre à Madame du Châtelet*, dedication at the beginning of *Alzire*, in *Oeuvres*, Vol II, p 456.

Maupertuis and Newton: the first significant French supporter of Newton's new physics was Pierre Varignon, who had begun the process of translating Newton's geometrical language into the more algorithmic Leibnizian calculus, in the 1690s. (See, for example, Boyer, p 435.) After that time, however, Continental scholars focused on the metaphysical nature of 'attraction', and little mathematical analysis was done specifically in support of the theory until Maupertuis.

Maupertuis, 'I do not at all wish to read': to Johann Bernouilli, 11 June 1731, quoted and translated in Terrall, p 58; Maupertuis, 'until we know what attraction is': to Johann Bernouilli, 23 April 1731, quoted and translated in ibid., p 57.

Voltaire to Maupertuis, 'Who would have thought', *Lettres Inédites*, Lettre II, 3 March 1733. Émilie to Maupertuis about lessons: for example, ibid., Lettre IV, January 1734. Note that evidence is scarce, but Zinsser, pp 62, 65, and Hamel, p 71, believe Émilie began her lessons with Maupertuis *before* she and Voltaire began courting, not after as I (and also Vaillot, p 82) have interpreted it.

Émilie and Maupertuis: most historians suggest they had an affair, but I agree with Zinsser, p 71, that this is not evident from their letters. Note, too, Zinsser suggests, pp 80–81, that Voltaire, Maupertuis, Algarotti and Clairaut were bisexual. Certainly Maupertuis and Clairaut lived together at times, and V had intense friendships with men such as Cideville and Thieriot before he met É; but, in a 1740 poem to (the openly homosexual) Frederick of Prussia, V said he had no interest in 'the affairs of Greece': *Correspondance*, Vol 2, pp 412–3.

Voltaire on holding back publishing his *Lettres Philosophiques*: see *Lettres de Voltaire et de sa célèbre amie*.

John Law had become a French citizen and, in 1720, the duc d'Orléans had promoted him to the position of France's comptroller-general of finance. (*Chambers Biographical Dictionary*.)

Émilie's passion for Voltaire: É to Sade, *Lettres de Voltaire et de sa célèbre amie*, 12 May 1734. Burning ('condemning') Voltaire: É to Maupertuis, *Lettres Inédites*, Lettre IX, 22 May 1734. É to Maupertuis on studying: ibid., Lettre X, 7 June 1734.

Voltaire, 'Is it only in England . . . ?': quoted in Faivre, p 26. Émilie in praise of the English: her own preface to *The Fable of the Bees*.

Newton's Arianism, etc.: Westfall, pp 314 ff, 333. Note that Cambridge also maintained two 'exempt' positions, but they were already occupied when Newton applied.

Voltaire's incisive critique: Cohen (1999), p 154; Maupertuis had used the same argument: Terrall, p 72.

On scholarly code of 'civility': Terrall, pp 3–4, 70; Maupertuis, 'It is not for me': from his *Discours sur les figures des astres* . . . , quoted and translated in ibid., p 70. Voltaire's style shocking, and making Newton seem subversive: ibid., pp 83–4. While Maupertuis had certainly been cautious about declaring his own support for 'attraction', his diplomatic style meant he was never ostracised in the Academy, as Terrall notes, p 84.

Empiricism as an evolving idea in the seventeenth century: see, for example, Cope, in Leitz and Cope (eds), pp 109–13. Cope mentions embarrassing empirical preoccupations with witchcraft, p 111, as does Barbara M. Benedict, ibid., pp 70–1, 96.

On Voltaire's *Philosophical Letters*: the original English edition was a bestseller and appeared in fifteen British editions during the eighteenth century; the French edition was banned, and would only appear in France as piecemeal articles: see introduction by Nicholas Cronk in recent English version given in Bibliography. Cronk says Voltaire composed his *Lettres* in 'such a way that his plea for toleration hit home on both sides of the Channel', so that 'the consequent complexity of the irony in the work is unprecedented'. Leonard Tancock also suggests the *Lettres* is 'almost as much a satire on the English' as on the French, perhaps as a strategy for protecting himself against his countrymen: Tancock, quoted in Johnson and Chandrasekar, Part I, p 431. But Cronk points out there were dissensions between Whigs and Tories in Britain, and that the new form of government had teething troubles that Voltaire satirised. Note, too, that fifteen years after V's *Philosophical Letters*, his countryman and contemporary Charles de Montesquieu later did much to spread the idea of constitutional monarchy throughout Europe, in his 1748 *L'esprit des lois* (*Spirit of the laws*). Note, too, that although Voltaire generally opposed nationalism, in the 1720s he had actually contributed to negative national stereotyping in which the British were seen as 'fierce' because they supposedly ate their meat rare: see George Sebastian Rousseau, in Leitz and Cope (eds), p 226.

*Philosophical Letters* today: in his preface, Jean-Luc Faivre calls it a 'tonic' for jaded modern humanists.

Quotes from *Philosophical Letters*: these are my translations from the longer French version, *Lettres Philosophiques*.

Locke and Newton's friendship: see Cohen (1999), p 38; Westfall, pp 470–1; Johnson and Chandrasekar, Part I, pp 445–6.

Locke on the soul: Locke, *An Essay Concerning Human Understanding*, in Cahn (ed.), p 622 (Voltaire's rendering of this passage: *Lettres Philosophiques*, p 84). On rejection of innate ideas: Locke, op. cit., pp 619–20 (Voltaire, op. cit., p 84). On whether purely material beings can think: Locke, op. cit., p 686 (Voltaire, op. cit., p 85). On the different provinces of faith and reason: Locke, op. cit., pp 703–12; in particular, even divine revelation must conform to reason: ibid., pp 704–5. Voltaire, 'ordered to believe': Voltaire, op. cit., p 85; Voltaire on 'burning'/'overthrowing religion': ibid., pp 84–5.

Voltaire to Argental, 'I wish I had said more': quoted in Faivre, p 28.

On Voltaire's apology via Madame de Richelieu, and the order of house arrest: Zinsser, p 92.

On the Cirey estate and its income: Zinsser, pp 14, 48. On Cirey's isolation: ibid., pp 107–9. On the marquis's military income and the military campaigns he served in: ibid., pp 48–9. Voltaire's investments, lottery: ibid., p 97.

Émilie to Sade, on Maupertuis leaving, going to Cirey, etc.: *Lettres de Voltaire et de sa célèbre amie*, undated, p 34.

Voltaire, 'Two hundred boxes': quoted in Zinsser, p 93; the journey to Cirey taking four to five days: ibid., p 108. 'Madame du Châtelet laughs': quoted in Hamel, p 66. Émilie to Maupertuis: *Lettres Inédites*, Lettres XII–XIV; É's letter regarding Héloise is Lettre XIII. Incidentally, É maintained certain Catholic rituals, including mass, at Cirey; this is mentioned by Graffigny (who noted that V only watched from a distance): quoted in Vaillot, p 168. See also ibid., p 123, for É acting out of prudence and family duty. On V's Catholicism: Zinsser, p 238; but see Faivre, p 26, for V's poem of 1732, in which he says, 'I am not a Christian,' and attacks much Christian dogma. É, too, attacked Christian dogma in her massive manuscript *Examens de la Bible*, refuting much of scripture by applying reason: Zinsser, pp 212–16 and Ehrman, pp 82 ff.

Clairaut, mathematics prodigy: Boyer, p 452. Clairaut's first lessons with Émilie (and probably Mlle de Thil): É to Maupertuis, *Lettres Inédites*, Lettre XI, 1734 (no exact date); see also Zinsser, p 73. Clairaut on É as 'altogether remarkable': Zinsser, p 75 (her source is I.O. Wade). Clairaut's easygoing character and taste for good living: Hamel, p 81.

Émilie to Sade on Voltaire's return: *Lettres de Voltaire et de sa célèbre amie*, 3 April 1735.

**CHAPTER 4: Émilie and Voltaire's Academy of Free Thought**

First canto of Voltaire's *La Pucelle* available in English translation from Jeanne d'Arc website at http://www.jeanne-darc.dk/p_multimedia/literature/0_voltaire/00_voltaire_contents.html. This excellent website also contains an English translation of V's article on Jeanne d'Arc, from the *Philosophical Dictionary*, 1752.

On the illusion that Émilie and her husband were merely Voltaire's protectors: Zinsser, pp 95–6, 133.

Émilie to Richelieu, four letters from 20 May to 15 June 1735: quoted and translated in Hamel, pp 60–2, and in Zinsser, pp 98–9. Cirey as 'earthly paradise': É to Richelieu, ibid., p 99.

Bored, unhappy women taking drugs, etc.: Badinter, p 155. Fontenelle on women hiding their minds and their feelings: quoted in Zinsser, p 101. Fontenelle on separating faith and reason: Wade, p 205. Fontenelle's *Entretiens* (*Conversations*) like those of elegant salons: online *Encarta* encyclopedia at http://fr.encarta.msn.com.

Madame de Graffigny on Émilie and Voltaire in love, same spoon: quoted and translated in Hamel, p 187. Chevalier de Villefort on Cirey: ibid., p 145. Algarotti on Cirey (and V on Algarotti, and on the 'voluptuous philosophers'): ibid., pp 141–2 (with correction by Peter Hambly). Graffigny studying Locke: Zinsser, p 120. Graffigny's description of her first supper at Cirey: Hamel, pp 170, 174–6.

Graffigny's descriptions of entertainments at Cirey, and of Émilie and Voltaire's work routine: Graffigny to Devaux, *Lettres du Dix-Huitième Siècle*, pp 191–4.

Graffigny on Émilie's conversation and dress: quoted and translated in Hamel, p 169; see also Anne Soprani's introduction to *Lettres d'Amour au Marquis de Saint-Lambert*, and Badinter, p 79 (from Asse [ed.], *Correspondance de Madame de*

*Graffigny*). On Voltaire wearing his wig: Hamel, p 170; É dressing formally at Cirey soirées: Zinsser, p 124.

Voltaire, 'When Émilie is ill': quoted in Zinsser, p 141; Émilie, 'Voltaire is my guide': quoted in ibid., p 142. É on V as a 'universal man', etc.: her own preface to *The Fable of the Bees*.

On the genesis and history of Mandeville's *Fable*: *The Cambridge Guide to Literature in English*, p 597. On differences with Émilie's version: Ehrman, pp 58 ff. *Fable* banned in France: Zinsser, p 25, Ehrman, p 59. For Émilie's translation and commentary of *The Fable of the Bees*: I am indebted to the online Women in Science project (headed by Judith Zinsser) for making available a digital version of the manuscript, as published in Ira O. Wade's *Studies on Voltaire* (1947). Here as elsewhere in the text, some of my translations are shortened or slightly paraphrased, although I have tried to keep the truth of the words as I interpret them. This is the approach É herself used in translating.

Before Mandeville, Locke wrote on education to avoid wrong ideas of shame, virtue, etc.: Postman, p 118. Locke on the social importance of disgrace and esteem: see Locke's *An Essay Concerning Human Understanding*, Book 2, Chapter 28, paragraph 11. See Wade (1971), p 518, for Locke's pivotal role in making the discussion of virtue and vice one of ethics rather than religion. Wade notes that Locke broadened the concept of morality beyond good and evil to include social relationships and obligations, not just the relationship between 'man' and God; thus he 'transformed a religious concept to a truly ethical one', and he also established rational rules for ethical behaviour (based on natural law and social contract).

Émilie reading/translating out loud: Vaillot, p 168. Graffigny, 'What a woman': Vaillot, pp 168–9, and Hamel, p 187.

On Graffigny and the double standard: Von Kulessa, especially pp 212–4. On Graffigny's life: Zinsser, pp 138–9, also Hamel. Success of Graffigny's novel *Lettres d'une Peruvienne*: Modern Language Association website at http://www.mla.org/store/CID43/PID147, regarding their 1993 edition of the novel.

Émilie to Sade [?], 'Newtonising': É to unnamed friend (É almost never addressed or signed her letters!), *Lettres de Voltaire et de sa célèbre amie*, 3 January 1736.

Some of Émilie's words used in *Elements*: Zinsser, p 149. Voltaire not embarking on *Elements* without É: Johnson and Chandrasekar, Part I, p 450. V, 'straining my brain with Newton'/'devil of a Newton': quoted and translated in Zinsser, p 148. See also Wade (1947), quoted in Badinter, p 282, about É's role in educating V for *Elements*.

On popularisations of Newton in the early eighteenth century: Johnson and Chandrasekar, Part II, pp 537 ff.

Descartes's philosophy: this is from his *Meditations in First Philosophy*; for example, Cahn (ed.), pp 411, 428–33, 435.

Kepler's law of equal areas: Leibniz claimed to have developed a vortex theory of planetary motion in which he used Kepler's second law in a different way from

Newton – although he was moved to publish his 'theory' *after* having read about Newton's approach: Cohen (1971), pp 152–3.

Newton disproving vortices using Kepler's third law: *Principia*, Book 2, scholium to Proposition 52. For this result in modern mathematical terms, see Resnick and Halliday, p 460, problem 19. Vortices inconsistent with (i) Kepler's second law, (ii) comets: *Principia*, scholium to Proposition 53, General Scholium, respectively. The *Elements* on vortex proofs: Johnson and Chandrasekar, Part II, p 534. Modern criticism of Newton's vortex proof: Neményi, in Cohen (1999), pp 187–8, but see also Smith, ibid., pp 188–9, 193.

**CHAPTER 5: Testing Newton: the 'New Argonauts'**

Émilie to Maupertuis: playing *Alzire* at Cirey, *Lettres Inédites*, Lettre XXII; going to the pole, ibid., Lettre XVI; praising the Lapp women, ibid., Lettre XXI.

Fontenelle on 'the new Argonauts': quoted in Terrall, p 94; Mitad del Mundo in Ecuador: http://www.histdoc.net/history/maupertu.html.

Maupertuis distancing himself from Newton's theory of gravity: Terrall, pp 95–9; focusing on the technical side of the journey: ibid., p 115. A detailed critical summary of Maupertuis's mathematical arguments on the shape of the earth is given by Todhunter, Articles 131 ff (for example, minimising error in triangulation/optimum number of triangles: Article 141). For Newton's, Huygens's and Clairaut's work on this matter, see also Todhunter.

Émilie to Richelieu on Maupertuis's unhappiness: 15 June 1735, quoted and translated in Terrall, p 101. Maupertuis, 'This voyage would hardly suit me': to Bernouilli, December 1735, ibid., p 101.

Maupertuis, 'how many ships have perished?': from Maupertuis's post-Arctic account, excerpted and translated in Fauvel and Gray (eds), p 455.

Astronomical definition of a day: for a detailed explanation and definitions, see Misner, Thorne and Wheeler, pp 23–4.

Newton on the shape of the earth: *Principia*, Book 3, Propositions 18–20. But see Cohen (1999), p 350: in some of his more complex applications of gravitational theory, Newton sometimes used brilliant hunches that were far from obvious, and which he occasionally did not properly prove. I have drawn on Maupertuis's summary (in his 1732 *Discourse*) of Newton's proof: it gives the essence rather than the exact letter of Newton's complex analysis.

Newton and Huygens, independence on circular motion law: Cohen (1999), p 115. Huygens naming 'centrifugal' force: ibid., p 82. Newton acknowledging Huygens: *Principia*, Book 1, scholium to Proposition 4. Note that in Book 1, Proposition 4, Newton revised his proof for centrifugal acceleration in terms of *centripetal* rather than centrifugal force (that is, reversing sign). Centrifugal force 'pseudo', including push-pull force electrical: Resnick and Halliday, pp 120–1, 657. Inertial forces: D'Inverno, pp 121–5, 130.

Huygens and pendulum/centrifugal force argument pre-dating Newton: Cohen (1999), p 349. Huygens on shape of earth post-dating Newton: Todhunter, p 35, Article 65. Huygens's use of Newton's 'channel' argument: ibid., p 30, Article 54. Huygens's theory of gravity, ibid., pp 33–4, Articles 59–60. Huygens's neo-Cartesian form of gravity (Huygens had been convinced by the *Principia* to abandon the idea of vortices, but he still did not accept action-at-a-distance): Cohen (1971), p 153, footnote 4.

Émilie on all matter having gravity: É's commentary on the *Principia*, Chapter 2, paragraph XX.

'Superb' theorems: the title is due to J.W.L. Glaisher, who applied it to the *Principia*, Book 1, Proposition 70. S. Chandrasekhar, pp 269 ff, extended this term to cover Propositions 70–5. Note that the theorems for spheres assume homogeneity. Maupertuis confused over 'superb' theorems: Terrall, p 56.

Measuring a degree of longitude, La Condamine's troubles: Todhunter, p 238. Maupertuis on the journey up the river valley, insects, reindeer herders, cataracts, etc.: from his *Figure de la terre*, quoted and translated in Terrall, pp 119–21. Maupertuis on the Arctic twilight: quoted and translated in Todhunter, p 97. Maupertuis returned hero: Terrall, p 118.

Maupertuis's conclusion about the flattened earth: from his *Figure de la terre*, excerpted and translated in Fauvel and Gray (eds), p 455. Clairaut extending Newton's calculations on flattening of the earth: Todhunter, Article 30, p 19, Article 37, p 23. Today's figure for the flattening of the earth: Cohen (1999), p 237. Earth is slightly pear-shaped: Fowles, p 138.

**CHAPTER 6: The danger in Newton: life, love and politics**

Voltaire on Descartes's philosophy: *Lettres Philosophiques*, 14th letter. Descartes's own words, for example, on body as an extended thing rather than a thing that thinks: *Meditations*, in Cahn (ed.), p 439.

Voltaire embroidering/circulating *Pucelle*: *Correspondance*, Vol 2, pp 610, 659, 991, and Bodanis, p 115.

Voltaire's *Le Mondain* is available online at http://www.udel.edu/braun/poetry/voltaire.html. I have also benefited from an interesting analysis/study guide of the poem at http://bacfrancais.chez.com/mondain.html.

On the *Le Mondain* furore: title meaning 'The Sophisticate': Zinsser, p 121; Émilie accompanying Voltaire on his escape: ibid., p 132; Voltaire on the 'horrible situation': quoted and translated in Ehrman, p 32. Argental as V's school friend: Zinsser, p 83. Cirey as 'earthly paradise': for example, V to the Richelieus, 12 January 1739. Hénault on Cirey: quoted in Faivre, p 31.

Émilie to Argental, on necessity of abandoning Cirey, etc.: December 1734, January 1735, BnF Document Number Z 15192. These dates should be 1736 and 1737, respectively: see Zinsser, pp 134–7. On the dangerous metaphysical chapter of *Elements* being the one concerning matter's ability to 'think': Zinsser, p 136. Émilie's help with *Alzire*: for historical details, ibid., p 82.

Voltaire's literary status in France today: see the following online articles from 2010: 'Académie Française: Le concours sur Voltaire' at http://fr.wikisource.org/wiki/Académie_française._-_Le_concours_sur_Voltaire, and 'Voltaire, homme de théâtre' at http://lire.ish-lyon.cnrs.fr/spip.php?article393.

Mary Somerville on 'artificial style of French tragedy': *Recollections*, original edition, p 111. Voltaire on Shakespeare: *Lettres Philosophiques*, beginning of 18th letter (on tragedy).

Émilie to Maupertuis, on *La Figure* and *Elements* being held up: *Lettres Inédites*, Lettres XXIII–XXV.

Maupertuis's Arctic publication being held up by opponents: Terrall, p 138. On the public spat between Cassini and Maupertuis's team: ibid., pp 134–9; see also Todhunter, Article 145, pp 74–5, and O'Connor and Robertson's biography of Maupertuis at http://www-groups.dcs.st-and.ac.uk/~history/Biographies/Maupertuis.html. Note that the Arctic team's credibility would be further attacked when two young women from Tornio turned up in Paris, a year after the team had returned: Terrall, pp 142 ff.

Newton sending Fontenelle his *Opticks*: Westfall, p 792.

Émilie and Voltaire on misplaced nationalism: É in *Institutions*, preface, paragraph VII; V in *Eléments*, pp 123–4. Algarotti, 'our Bolognese savante', etc.: *Newtonianism for the Ladies*, my translations from the 1738 French version (translated from the Italian by M. Duperron de Castera), available online from the BnF's Gallica 2 catalogue. Note that references to É's *Institutions* are from the 1740 edition at http://athena.lib.muohio.edu/djvu/index.php?file=institutions.djvu.

Analysis and history of Algarotti's *Newtonianism for the Ladies*: Mazzotti (including, for example, the book's successive revisions in light of Church criticism, pp 25–6, 30 and footnote 31 on historian Mauro De Zan). Positive effect of Algarotti's book on women and other lay readers: Mazzotti, Findlen (2003); craze for English tea and hats: Mazzotti, p 15. Mazzotti's excellent article is available at http://www.cis.unibo.it/cis13b/bsco3/algarotti/introbyed/algintrobyed.pdf. Émilie's interest in science was serious, most women's merely fashionable: Vaillot, p 19. On the erotic nature of Algarotti's book: Émilie to Richelieu, *Lettres de Voltaire et de sa célèbre amie*; the seductiveness of mathematicians: Algarotti, pp 50–1.

On the conversational style of Fontenelle and Algarotti: of course, Plato had used this style, in his Socratic Dialogues, and Galileo used it in his pro-Copernican *Dialogue Concerning the Two Chief World Systems*, which led to his being charged with heresy in 1634. But fifty years later, in his *Conversations*, Fontenelle was the first to introduce a female protagonist, and a seductive style – a style that Algarotti used another half-century later: Findlen (1993).

Voltaire angry at addition of 'for everybody' to title of *Elements*: Émilie to Maupertuis, *Lettres Inédites*, Letter XXIV, 9 May 1738. É to Argental, hoping V's restraint in publishing *Elements* would appease censor regarding unauthorised Dutch edition: Besterman, letter number 123, May 1738.

Émilie to Richelieu, on Algarotti (and on Fontenelle being 'neither a woman nor a Newtonian'): *Lettres de Voltaire et de sa célèbre amie*, apparently 17 February. É to Maupertuis, on Algarotti: *Lettres Inédites*, Lettre XXIV, 9 May 1738. É, 'sixty-four times less': quoted and translated in Ehrman, p 29.

Voltaire to Argenson, on Lenglet imprisoned, Cartesian daydreams: 8 August 1743, *Correspondance*, Vol 2, p 667. V to Ffolkes, on being a martyr: ibid., 25 November 1743, p 721.

On Laura Bassi: Kleinert, Findlen (1993 and 2005).

On Voltaire's Academy memberships: V to Laura Bassi: *Correspondance* Vol 2, pp 821, 834; V's election to Royal Society: V to Ffolkes: ibid., p 721. See also V to D'Alion: ibid., p 868, for list of Academy memberships V had obtained by mid-1745: London, Edinburgh, Berlin and Bologna. Many more elections would follow. Note, however, that V's election to the Académie Française was surprisingly slow given his literary success: Zinsser, pp 235–9 (including Tencin against V, pp 235–6); even the non-literary Maupertuis and Mairan were members before Voltaire: Terrall, pp 196–7. 'All Paris studies Newton': quoted in Johnson and Chandrasekhar, Part II, p 537.

## CHAPTER 7: The nature of light: Émilie takes on Newton

Émilie studying *Opticks* at Cirey, doing experiments on light and fire: Zinsser, pp 153–62.

Newton, Hooke, Boyle and Descartes prism experiments: Rankin, p 92, also Westfall, p 164. Note that Newton projected his spectrum onto a wall 22 feet away in his 1672 experiment, and in his later *Opticks*, he mentioned a similar distance of 18½ feet. Newton and the gold leaf: Westfall, p 163. Newton on 'spectrum': see Cohen's preface to *Opticks.*

Seven colours/octave: *Opticks*, and online article 'Music for Measure: On the 300th Anniversary of Newton's "Opticks" ', at http://home.vicnet.net.au/~colmusic/. 'Families' of colour gradations: see, for example, Resnick and Halliday, p 994, Figure 40.2.

Newton's experiments on the immutability of the spectral colours: *Opticks*, Part I, Book I, Experiment 12 (Proposition V), and Part II, Book I, Experiment 5 (Proposition II). The original spectrum experiment: *ibid.*, Part I, Book I, Experiment 3.

On criticisms of Newton's experiments on colour, Hooke's and Huygens's reactions: Westfall, pp 246–50. It took four years of arguing to convince the Royal Society to confirm Newton's spectrum experiment: Rankin, p 102, Westfall, p 274. On

Algarotti's light experiments, Rizzetti, 'British prisms', empiricism, Newton replicating his experiments for Italians, etc.: Mazzotti, pp 18–25. Newton replicating experiments regarding Rizzetti, Mariotte: Westfall, pp 795–6. Émilie to Maupertuis, on Newton's colour experiments and critics: Besterman, letter number 152, December 1738. In this letter, Émilie also mentioned an earlier (1728) criticism of Castel, saying, 'Newton had refuted this objection 50 years ago.'

Telescopes: Newton's description of his telescope-making method: *Opticks*, pp 103–5. Gregory's reflecting design: Stillwell, p 131. Lippershey telescopes: *Oxford Reference Encyclopedia*; Snell (and also Descartes) on refraction: Resnick and Halliday, p 1015. Newton's one-inch mirror, and reflecting telescopes today: *How Is it Done?*, pp 181–2. Newton's telescope six inches, etc.: Rankin, p 97.

The physics of reflection and refraction: see, for example, Resnick and Halliday. On Euclid, ibn-al Haitham: Boyer, pp 173, 240.

Newton's first draft of *Opticks* being destroyed: Rankin, p 104, Westfall, p 277. Newton as 'slave to Philosophy': Newton to Oldenburg, Secretary of the Royal Society, quoted in Westfall, pp 275–6.

On sales of the *Principia*: Whiteside, in Cohen (1999), p 22, footnote 39. Halley's heroic effort to ensure the *Principia* published: Stillwell, p 115.

Voltaire spending several million dollars on his laboratory: Bodanis, p 116. V, 'one cannot be learned without money': quoted and translated in Zinsser, p 153.

Using the forges at Cirey: Zinsser, p 155, Bodanis, p 118. Voltaire using two thousand pounds of iron: *Correspondance*, Vol 2, p 120. Glass containers: Bodanis, p 118.

Newton on light: it was in 'Query 29' that he proposed light was made of 'small bodies' which travel in straight lines without bending into shadows.

Newton's particle explanation of refraction: Voltaire, *Eléments*, pp 97–102. Newton's approach purely mathematical: see, for example, *Opticks*, p 80. Newton, 'not arguing at all about the nature of the rays': *Principia*, Book 1, scholium to Proposition 96. See also Cohen's preface to *Opticks* (Newton versus Huygens), pp xlvi–xlvii. Newton on speed of light greater in water: implied by *Principia*, Book 1, Proposition 95. Newton's mathematical proofs on 'attraction' as a cause of refraction: ibid., Book 1, Propositions 94–8. Émilie/Clairaut on Newton's 'refraction' theorems: É's commentary on the *Principia*, p 189. For Huygens's wave theory calculations on why the speed of light should slow down in denser media: see, for example, Resnick and Halliday, pp 1021–3.

Émilie having only one month to write her essay on fire: É to Maupertuis, *Lettres Inédites*, Lettre XXVII; on secrecy, etc.: ibid., Lettres XXVI–XXVII.

Seguin, pp 334–5, recently made an extensive search for early women's contributions to the Paris Academy of Sciences, and has found only Émilie's.

Émilie's experiments and conclusions are from the revised, 140-page, 1744 version of her *Dissertation sur . . . feu*: on the relationship between colour and heat, pp 69–70,

137; on the 'heaviness' or otherwise of fire, Chapter VI, pp 25 ff. Newton on total internal reflection: *Opticks*, Book 1, Experiment 19; for a modern illustration: Resnick and Halliday, pp 1024–5. Evaporation, kinetic energy and temperature: ibid., pp 580, 604. É on evaporation: *Dissertation sur . . . feu*, pp 54, 69.

On Herschel's experiment: for an accessible detailed description: Institute of Physics website at http://www.practicalphysics.org; for a contemporary account: Mary Somerville, *Connexion*, p 232. See also Bodanis, p 124. On Rochon: Badinter, p 307.

Voltaire, 'Fire is matter': *Eléments*, Chapter IX, quoted and translated in Johnson and Chandrasekar, p 532.

Voltaire on his own and Émilie's essays: V to d'Argens, *Correspondance*, Vol 2, pp 120; V to Alleurs, ibid., p 187. V's approach to Academy, and precedent of publishing runners-up: Zinsser, p 169.

Other winners of essay on fire of no consequence: *The Biographical Dictionary of Scientists.*

**CHAPTER 8: Searching for 'energy': Émilie discovers Leibniz**

Benjamin Thompson and heat as motion not substance: Resnick and Halliday, pp 545–6.

Leibniz on *vis viva*: extracted in Wiener (ed.), pp 181–4, including original walking example, p 183; Leibniz using different terms for 'living force': ibid., pp 125, 135, 156, 183. Leibniz on momentum as 'impetus': ibid., p 123. Émilie on 'living force': *Institutions*, Chapter 21, especially p 428 (experimental evidence), pp 433–4 (the two walkers/travellers).

Émilie's calculations on the 'force' (energy) of light, using the formula $mv^2$: *Dissertation sur . . . feu*, pp 31 ff.

Émilie to Maupertuis, on deleting reference to Mairan and on domesticity/ 'If I were a man': *Lettres Inédites*, Lettres XXVIII and XXIX, October and December 1738; Besterman, letter number 152, December 1738; also Badinter, p 333, Zinsser, pp 112–15.

Playfair on Émilie: quoted in Todhunter, p 363.

Desfontaines on Voltaire (*Voltairomanie*): quoted in Vaillot, p 174.

Frederick on Émilie's essay: quoted and translated in Hamel, pp 162–3.

Brussels, Voltaire on leaving Cirey: 1 January 1739, *Correspondance*, Vol 2, pp 2–3. Émilie's 'power of attorney', taking charge of the case: Zinsser, p 49.

Voltaire to Helvétius on monads, etc.: *Correspondance*, Vol 2, p 222.

Leibniz on monads 'expressing the whole universe': *Discourse on Metaphysics* (1686), section IX, extracted in Wiener (ed.), pp 300–1; see also *The Monadology* (1714), extracted in ibid., pp 533–52.

Émilie's explication of the philosophy underpinning monads, including the analogy of the watch (my paraphrasing): *Institutions*, pp 132–8.

Anne Conway, monads of 'vital active force', Ragley Hall: Merchant, pp 254, 264–8. Leibniz on Conway (he calls her the Countess of Connaway): *New Essays Concerning Human Understanding*, p 67.

On Leibniz's logic/free will: George Sher's introduction to Cahn (ed.), pp 573–4.

Voltaire's objection to monads: a decade later, Euler would publish a similar objection, that it was ridiculous to describe materiality in terms of immaterial monads: Terrall, p 259. Leibniz's narrow 'logical base': George Sher's introduction to Cahn (ed.), p 573.

Émilie rebutting Locke/matter thinking: *Institutions*, pp 65–6. No sufficient reason for attraction: ibid., pp 327–9, 333. Gravity not inherent in matter: ibid., pp 330–2. Leibniz on sufficient reason: extracted in Wiener (ed.), p 94. Leibniz refuting Locke on possibility of material cause of thought: *New Essays Concerning Human Understanding*, p 428.

Descartes on soul as indivisible and therefore immortal: *Meditations on First Philosophy*, paragraphs 13 and 14, extracted in Cahn (ed.), pp 410–11. (Note that Descartes's 'proof' actually works better as a statement of conservation of mass, because he applied it not only to the soul, but also to all the other fundamental, non-composite 'substances created by God', 'substances' that modern physicists call 'elementary particles'.) Locke on circular arguments: see, for example, *An Essay Concerning Human Understanding*, Book II, Chapter 1, paragraph 10, in Cahn (ed.), p 622. Locke did believe thought was immaterial, but he recognised he could not prove this beyond a general definition of 'spiritual' in terms of our inner sense of ourselves (ibid., Book II, Chapter 23, paragraphs 15–16, in Cahn [ed.], p 652), because no-one knew 'wherein thinking consists', and so perhaps purely material beings can think (ibid., Book IV, Chapter 3, paragraph 6, in Cahn [ed.], p 686). Locke identified 'consciousness' rather than 'soul' as the defining quality of a person (ibid., Book II, Chapter 27, paragraphs 10 ff, in Cahn [ed.], pp 656 ff).

On absolute versus relative space and time: Einstein, 'only fruitful one', etc.: from his foreword to *Concepts of Space* (M. Jammer [ed.], Harper, New York, 1954), quoted in Misner, Thorne and Wheeler, p 19. On pp 23, 26, Misner et al. show why Newton's choice of absolute space and time was important, that is, in making 'motion look simple'; see also Cohen (1999). Émilie on absolute and relative frames: *Institutions*, pp 219–20; on space: ibid., for example, p 94. Leibniz on relative time and space: extracted in Wiener (ed.), p 164. On Kant and later developments (for example, by Whitehead) of link between mathematics and the brain: Kline, p 341.

For a good introductory account of the two-slit experiment and the meaning of the so-called 'wave-particle' duality, see Penrose, pp 299–304, or Davies and Gribbin, pp 203–6.

On metaphysics today: Randall's preprint is available at http://www.elea.org/. See also Moriarty, and Smith.

Émilie on 'necessary corollaries', 'marvellously well': *Institutions*, pp 317, 319.

Newton on force: his general definition in his second law of motion is now written $F = ma$ ($a$ = acceleration, $a$ is a second derivative). Newton did not use this formulation, although he certainly could have done as he invented calculus methods. Furthermore, Jacob Hermann expressed Newton's law as $F = ma$ in 1716 (see *Stanford Encyclopedia*), that is, *before* Newton prepared his third edition of the *Principia*. This seems to prove he was trying to minimise the introduction of new concepts, especially calculus-based ones, in order to make the gravitational theory itself seem more acceptable. In the 1740s, Euler began the modern trend to use $F = ma$ consistently.

## CHAPTER 9: Mathematics and free will

Émilie on cycloid, Bernouilli, etc.: *Institutions*, pp 364 ff, 370. Note that she also gave a detailed description of other mathematical properties of cycloids, including their role in pendulum clocks: ibid., pp 363–8.

Bernouilli and the brachistochrone challenge: I have drawn primarily from Chandrasekhar, pp 571–3, and Westfall, pp 581–3 (including historical evidence on Bernouilli and Leibniz's motivations). The quote from Bernouilli is translated in Westfall, p 582. On Newton's 'toadies': Westfall, p 721. On Newton understanding calculus foundations better than Leibniz: Boyer, p 404. Newton's calculus in geometrical style in the *Principia* (and aid to concrete thought): Cohen (1999), for example, pp 114–15, 122–7, and Chandrasekhar, for example, p 273.

Note that Hall, pp 105–6, claims that Bernouilli and Leibniz did not set out to trap Newton, but his evidence seems contradictory – especially since he presents Leibniz's claim that only 'his' differential calculus could yield a solution of the brachistochrone problem. Rickey specifically points out that Hall's argument on this point is flawed: see Rickey at http://www.math.usma.edu. However, Hall does point out the rivalry between Jean Bernouilli and his brother Jacques, who sent in a superior solution to Jean's own. Boyer, p 417, notes that Jean's original cycloid solution contained an error, and when he received Jacques's correct solution, he tried to claim it as his own. On the calculus priority dispute, note, too, that it seems Leibniz withheld the fact that he had seen an early draft of Newton's calculus quite some time before publishing his own; Leibniz's work is nevertheless held to be independent of Newton's – calculus was already 'in the air' – although Newton's disciples did not think so at the time, hence their accusation of plagiarism. For a fuller account, see Westfall, Chapter 14, especially p 776.

Newton's early calculus papers rejected for publication: Stillwell, p 109. Newton's solution to the brachistochrone problem: Chandrasekhar, pp 573–4, 577–8; see also Rickey, and Herrera.

Newton's long working hours: Whiteside, excerpted in Fauvel and Gray (eds), p 423. Newton at the Mint: see Rankin, pp 143–6, for a succinct account of the looming national bankruptcy crisis, which made counterfeiting treasonable; this is the context for the well-known fact that Newton pursued counterfeiters zealously, sending some to the gallows. For a fuller account, see Westfall.

Newton's alchemy: see Cohen (1999), pp 60–1, for excellent discussion of Newton's genius in separating the experimental from the hypothetical and mystical, and for further discussion (and references to other sources, notably Betty Jo Dobbs) on Newton's alchemy and religion. See also Westfall. Regarding alchemy, both Newton and Boyle were atomists; *Chambers Biographical Dictionary* says of Boyle, 'His alchemy was a logical outcome of his atomism. If every substance is merely a rearrangement of the same basic elements, transmutations should be possible. Modern atomic physics has proved him right.' The same could be said of Newton's account of hard atoms in 'Query 31'. For a popular account of the philosophical significance of Newton's 'Query 31' in terms of life force, etc., see, for example, Merchant, pp 283–7, Rankin, pp 108–10. For the role of alchemy in Newton's conception of force, see Cohen (1999), pp 58–64.

Newton's other surprising (to modern eyes) preoccupation was biblical prophecy: in his study, Newton actually developed scientific (astronomical) techniques for dating biblical events. His writings on the Bible are in general fascinating rather than fanatical (although his hatred of the Catholic Church's 'idolatry' and 'money-grabbing' ['selling relics'] was extreme).

The role of mysticism in Newton's science: two of the people who have studied Newton's papers most closely are John Maynard Keynes and Derek Whiteside. Keynes believed Newton saw the universe as God's cryptogram, and that his scientific work was motivated by the quest to solve this riddle rather than by a belief in the power of mathematics and empiricism. (See also Kline, p 59, for private letter from Newton to Bentley, 10 December 1692, in which Newton said, 'When I wrote [the *Principia*], I had an eye on such principles that might work for considering [educating?] men for belief in a Deity; and nothing can rejoice me more than to find it useful for that purpose.' Of course, Leibniz and many others held similar views, as my narrative has shown; see also Kline, p 60.) In terms of Newton's study of mathematics rather than natural philosophy, however, Whiteside could not find any hint of mysticism in any of the thousands of pages of mathematics in the Newton archive, leading him to conclude that Newton loved mathematics for its own sake, and that his pursuit of mathematics was quite separate from any such mysticism. See Keynes and Whiteside, excerpted and translated in Fauvel and Gray (eds), original documents 12.F4 and 12.F5, pp 421–3.

On calculus: for a more detailed exposition of the rise of mathematical rigour in calculus, see, for example, Boyer (1959), p 213 (differentials as monads), p 226. Boyer notes that the very earliest pioneers of calculus included Archimedes, al-Haitham, Oresme, Viète.

Leibniz on best possible world: *The Theodicy* (1710), extracted in Wiener (ed.), pp 509–22. Émilie on best possible world: *Institutions*, pp 49–50. For modern appeal of Leibnizian harmony, see, for example, Merchant.

Voltaire on best possible world: *Candide*, see, for example, end of Chapter 5 and beginning of Chapter 6. For Voltaire's moving poem on the Lisbon earthquake, and Rousseau's response to it (letter to V, 18 August 1756), see the edition of *Candide* by Hachette, pp 169–70. Rousseau pointed out that no philosopher would deny there are specific instances of evil, but these must be distinguished from a general belief that the world is full of suffering, which he believed was Voltaire's view. Indeed, Dynes suggests that, rather than seeing it as a sign of the inherent unfairness in the world, Rousseau looked for practical solutions (such as better architecture and town planning) to prevent such disasters. (*Candide* sold thirty thousand copies: Dynes, pp 14–15.) Other philosophers in support of Leibniz: for a detailed account of the different reactions of Voltaire and Diderot, see Delon, pp 39–43: Diderot was drawn to the elegance of the monad philosophy (which he extended to the concept of a 'collective unconscious'), while Voltaire, 'mistrustful of all systems' of thought, saw Leibniz's philosophy as simplistic.

Émilie discussing Mairan and refraction of light with her son: *Institutions*, p 371. Heron of Alexandria and shortest path of light: Boyer, p 174.

The König affair: Émilie imperious with König (she even called him a 'lackey'!): quoted and translated in Zinsser, p 190. König asking É to sign his lessons: Ehrman, p 34. Voltaire defended É against König: see, for example, V to Helvétius, 24 January 1740, and V to Bernouilli, 30 January 1740, *Correspondance*, Vol 2, pp 251, 255–7. É to Maupertuis about Bernouilli II as tutor: *Lettres Inédites*, Lettre XXXII; Maupertuis dissuading Bernouilli from tutoring É: Ehrman, p 35. V to Maupertuis: 21 July, 9 August 1740, *Correspondance*, Vol 2, pp 348–9, 354–5. É to Maupertuis: quoted and translated in Hamel, p 217.

On Émilie's children: Voltaire to 's-Gravesande about tutor for É's son: 29 February 1740, *Correspondance*, Vol 2, pp 270–1. Son at the Royal Musketeers: Zinsser, p 222. Graffigny on É's daughter: Graffigny to Devaux, December 1738, quoted and translated in Hamel, pp 182–3, and Zinsser, p 40. On daughter's 'advantageous' marriage: Voltaire, 4 April 1743, *Correspondance*, Vol 2, p 637. This letter also shows V's genuine affection for É's daughter, and his regret at her being 'married off', although he later wrote (to Marmontel, February 1748, *Correspondance*, Vol 2, p 1042) of the importance of É's providing the right connections for her family, including her daughter.

Public response to É's *Fundamentals*: favourably reviewed in *Journal des savants:* Zinsser, p 191. Deschamps: quoted and translated in Ehrman, p 55. Wolff: quoted and translated in Zinsser, p 210. Clairaut and others' responses: Zinsser, p 195. Cideville: quoted and translated in Hamel, p 212. Voltaire to Helvétius: *Correspondance*, Vol 2, p 424. Maupertuis: discussed and quoted in Badinter, pp 328–9. Frederick: discussed (including E's offering Frederick lessons) and quoted in Badinter, pp 326–7. É clearer than Wolff: Ehrman, p 47, says contemporary reviews spoke of her 'remarkably clear exposition of German philosophy'. On Wolff as Leibniz's disciple: Terrall, p 182. É to Frederick, 'French sauce': quoted and translated in Ehrman, p 47.

Leibniz on carping criticisms: *Essay on Dynamics* (1695), extracted in Wiener (ed.), p 121.

Mairan as critic and defender of French science: Zinsser, p 191.

Émilie to Argental on hating Frederick: quoted and translated in Ehrman, p 33. É also said, 'The king of Prussia is a very dangerous rival for me': É to Argental (or Sade, Richelieu?), *Lettres de Voltaire et de sa célèbre amie*, 28 June 1743.

Émilie and Voltaire disagreeing on 'living force': É to Argental, 'don't know if I will reply','greater conformity in all else': 1741, BNF Document Number Z 15192, p 224. V to Mairan: 12 March 1741, 25(?) March 1741, *Correspondance*, Vol 2, pp 452, 464. V to Cideville on trusting É: 13 March 1741, ibid., Vol 2, p 455.

Émilie versus Mairan: É on Mairan: *Institutions*, p 433. Mairan's ironical accusations: Badinter, p 330. E's response: *Réponse . . . de Mairan . . .*, p 3. Reviews of É's and Mairan's open letters: Zinsser, pp 204–5.

Émilie to Maupertuis on *vis viva*: Besterman, letter numbers 118, 120 ('dispute of words'), 122, 124, from February to May 1738. É to Maupertuis: *Lettres Inédites*, Lettre XXXV, 29 May 1741.

D'Alembert exploded the dichotomy in the *vis viva* debate in his 1743 treatise on mechanics: see, for example, Resnick and Halliday, p 222. Note that Bernouilli had found his derivation of $mv^2$ in an essay for the Academy of Sciences prize in 1724; Émilie thought he should have won the prize: É to Maupertuis, Besterman, letter number 120, February 1738. See also Terrall, '*Vis viva* revisited', for example, p 199. Note, too, that Leibniz had verbally defined momentum in terms of an integral of force (he called it 'impetus') with respect to time: extracted in Wiener (ed.), p 124. Émilie gave an analogous verbal definition: *Institutions*, p 415.

Émilie to Argental, on returning to Cirey: BNF document number Z 15192, letter numbers 62, 63. Outcome of inheritance case: Zinsser, pp 201–2 (cf dowry and regiment, p 48). See also Voltaire, *Correspondance*, Vol 2, pp 1015 (É's 'power of attorney'), pp 1055–8 (V to Hoensbroeck, buyer of the contested land).

Voltaire to Cideville, friendship not 'amour': *Correspondance*, Vol 2, pp 506–7.

Voltaire to Frederick (Émilie his king): for example, *Correspondance*, Vol 2, p 714 (to Champbonin), p 773 (to Vernet).

Voltaire to Frederick on poem on Joan, locked manuscripts: *Correspondance*, Vol 2, pp 610, 659, 991 (and note 4, p 1334). Émilie and Voltaire's fiery relationship: Hamel, p 180.

**CHAPTER 10: The re-emergence of Madame Newton du Châtelet**

Voltaire to Condamine: October 1744, *Correspondance*, Vol 2, p 814. V to Argenson: ibid., p 755. V to Richelieu: ibid., p 771. Argenson, school friend: Zinsser, p 83. *Chambers Biographical Dictionary* notes that Argenson's father had created the secret police force that would hound his son's friend Voltaire!

Jacquier at Cirey: Voltaire to Vernet, *Correspondance*, Vol 2, p 773. Italian and German translation of *Institutions*: Zinsser, p 209; Bassi using *Institutions*: ibid., p 210.

'Madame Newton-Pompom-du Châtelet': see, for example, Voltaire to Argental, January 1745, *Correspondance*, Vol 2, p 830; 'Madame Newton': see, for example, ibid., pp 863, 914, 1038. V to Algarotti, 'sacred abyss': ibid., p 877 (translated p 1282).

On the accuracy of Émilie's translation of the *Principia:* Cohen's preface to his and Whitman's *Principia*, pp xii–xiii, plus footnote on p 647. Other references to Émilie's accuracy: Cohen (1971), pp 25n, 67n, p 7 ('great' eighteenth-century editions). See also Chandrasekhar, footnote on p 71.

On the *Principia*'s logical structuring: I mentioned the Russian doll image that came to me when I was struggling to follow the logical order of Newton's proof of his theory of gravity, but Peterson, p 87, quotes Whiteside as saying that the *Principia's* 'logical structure is slipshod, the level of verbal fluency none too high, its arguments unnecessarily diffuse and repetitive, and its very content on occasion markedly irrelevant to its professed theme'. As Peterson says, however, in judging the *Principia* we should not forget that it was composed in an astonishingly short time, under Halley's urging: although Newton had been working on his ideas for two decades, Westfall, p 450, notes that he wrote much of the *Principia* between August 1685 and spring 1686; according to Stillwell, pp 114–15, all up it took about two and a half years. Not surprisingly, he made a few errors in the first edition, which rivals like Hooke and Bernouilli delighted in pointing out: Westfall pp 384–5, 742–3, and Cohen (1999), pp 168–171.

The various languages in which the *Principia* now appears: Cohen (1999), p 26.

Voltaire to Algarotti about Gabrielle-Pauline: *Correspondance*, Vol 2, pp 987, 995; V to Gabrielle-Pauline: ibid., p 939. (For French translations of these letters, which V wrote in Italian, see ibid., pp 1312, 1332, 1336.)

Émilie's election to the Academy of Bologna: É to Jacquier on using Bologna title (and on her son and daughter): 12 November and 17 December 1745, *Lettres inédites de la Marquise du Châtelet et la duchesse de Choiseul.* Clairaut to Jacquier: ibid., editorial footnote. É to Jacquier on encouraging women: 4 September 1746, quoted and translated in Zinsser, pp 209–10.

Émilie to Argental, 'cruel' situation: quoted in Mauzi's preface to the 1961 edition of É's *Discours sur le bonheur*, p xxix. É on the end of her passion with Voltaire: ibid., p 32. For the early history of the publication of the *Discours sur le bonheur*, including Florent-Louis's and others' reactions, see Mauzi's preface, pp cxx ff.

Voltaire to Denis: 27 December 1745, *Correspondance*, Vol 2, pp 923, 1300. See also ibid., for example, pp 840–1, 920–34, 1297–1306, 1054.

Voltaire to Algarotti, 'our immortal Émilie': *Correspondance*, Vol 2, p 1011.

Émilie to Maupertuis on 'superb' theorems: Besterman, letter number 126, May 1738.

Discovery of Émilie's manuscripts by Wade: Badinter, p 472, and Zinsser, p 280. Émilie's mathematical notebooks: Zinsser, p 275.

Émilie's commentary on the *Principia* is in Book 2 of her translation.

Émilie on Hooke and Newton on gravity: É's commentary on the *Principia*, p 6. See also Cohen (1999), pp 76–8, plus article by Nauenberg in ibid., pp 78–82, and Westfall, pp 382–7. Hooke 'hopelessly wrong': Curtis Wilson, quoted in Cohen (1999), p 78. Halley saving the *Principia* from Newton's tantrum: Stillwell, p 115.

Émilie on Newton's creation of theory of gravity: É's commentary on the *Principia*, Chapter 2, especially paragraphs I–XX; simpler version: *Institutions*, pp 297–301. Émilie on simplified proof, Proposition 70: *Institutions*, p 327.

On Newton correcting for the influence of other planets (three-body problem): *Principia*, Book 1, Proposition 66 and its corollaries; for a detailed commentary, see Cohen (1999), pp 155 ff and Chapter 8, and see also Émilie's commentary on the *Principia*, Chapter 2, sections XVII, XL–XLIII.

Clairaut rejecting Newton's theory: Clairaut, Academy paper, November 1747, and retraction, May 1749, excerpted and translated in Fauvel and Gray (eds), pp 458–9. Error in his calculations: Peterson, p 136. Euler's and D'Alembert's proposed changes to Newton: excerpted and translated in Fauvel and Gray (eds), pp 454, 456–7. Clairaut to Euler on vortices: excerpted and translated in ibid., p 457. D'Alembert not wanting to overthrow Newton: quoted and translated in Peterson, pp 134–5. Euler to Clairaut on the latter's definitive proof of inverse-square law: quoted and translated in Cohen (1999), p 204.

Moving to Lunéville, ancestral region: Zinsser, pp 248, 258. Voltaire to Argental, 'small planet': 1 February 1748, *Correspondance*, Vol 2, pp 1041–2; V to Denis, 'My dear, here I am': ibid., p 1040 (French translation, p 1351). Émilie singing *Issé*: V to Madame d'Argental, ibid., p 1050.

Émilie meeting Saint-Lambert: my description is drawn from Anne Soprani's introduction to *Lettres d'Amour au Marquis de Saint-Lambert*.

**CHAPTER 11: Love letters to Saint-Lambert**

Early letters from Émilie to Saint-Lambert: see, for example, *Lettres d'Amour*, letter numbers 5, 6, 11, 12, 18, 21, 22 ('my Newton'). Quotations from É to St-Lambert are from Soprani edition of É's *Lettres D'Amour* unless otherwise noted.

Voltaire defending Émilie as scholar, 'neglecting her family': V to Marmontel, 2 February 1748, *Correspondance*, Vol 2, p 1042. See also ibid., pp 619–29, for V defending É against another satire against her.

Voltaire's response to Saint-Lambert: Émilie to Saint-Lambert regarding V, 'no illusion'/'blush at having loved him': quoted and translated in Zinsser, p 269. Voltaire still jealous of Clairaut: Hamel, pp 78–9, and Vaillot, pp 304–5 (both are based on Longchamps's memoir; however, this memoir was written long after the events, and is now regarded as historically inaccurate, especially in its more sensational details: see Soprani, p 265).

The jealous gods: Voltaire (*Contes en vers: Oeuvres Complets*, Vol 10, Flammarion, 1877, p 541), quoted in Zinsser, pp 260–1. Denis on Émilie: quoted and translated in Erhman, p 31.

Denis jealous of Émilie ('that woman'): Vaillot, p 312, and Mercier, p 319. É and V talking into the night: Soprani, pp 23–4. V and É reconciled: see, for example, V to Fawkener: *Correspondance*, Vol 2, p 1106. É and V at Cirey for Christmas: ibid., pp 1116–18.

D'Alembert abandoned by Tencin: Boyer, p 447. Nunnery: Zinsser, p 271.

Émilie to Saint-Lambert, 'thousand dagger blows': 28 March 1749, *Lettres d'Amour*, letter number 74. Note there is an often-told story about Émilie and Voltaire luring her husband to Cirey in late January in order to deceive him into thinking the baby was his: see, for example, Mercier, p 304, based on later writings by contemporaries Longchamps and Collé. This story seems unlikely, given the gossips already knew of her relationship with Saint-Lambert; see also Zinsser, p 271, footnote 65.)

Émilie to Saint-Lambert on anguish, including no longer believing the truth of mathematics: *Lettres d'Amour*, letter number 74.

Émilie resuming the *Principia*, 'terrible work'/'constitution of iron'/'Do not reproach me my Newton'/reassuring Saint-Lambert about not socialising: quoted in Mercier, p 317, and Zinsser, p 273. Voltaire on É dividing nine digits in her head: quoted in Badinter, p 200.

Émilie to Saint-Lambert, on being abstemious, well, baby moving: quoted in Mercier, p 317, Soprani, p 25.

Voltaire to Frederick on not leaving Émilie: quoted in Mercier, p 309. V to Frederick, 'has still not delivered', etc.: *Correspondance*, Vol 3, p 96.

Émilie's *Principia* to librarian: É to Sallier, *Lettres d'Amour*, p 244.

Émilie to Saint-Lambert, 'my belly has dropped'/'so sad this evening': *Lettres d'Amour*, letter number 99.

Voltaire on the birth of Émilie's baby: see, for example, to Argenson, *Correspondance*, Vol 3, p 99.

Émilie's death: from Longchamps's memoir, summarised in Mercier, pp 333–5, Vaillot, pp 315–16, and Zinsser, p 278. Recall (see earlier note) that Longchamps's memoir was written long after the event, and was to some extent sensationalised.

**CHAPTER 12: Mourning Émilie**

Saint-Lambert grieving, breakdown: Zinsser, pp 287–8 (including Graffigny quote).

Voltaire on Émilie's death: September to November 1749, *Correspondance*, Vol 3, especially pp 104–5, 106–7, 108, 109–10, 115, 121, 122, 124.

Émilie quickly forgotten: of her obituaries, only Voltaire's and Maupertuis's were truly supportive: Badinter, p 476. Zinsser, p 281, and Mercier, p 340. Note that many mocked Émilie's death, including the novelist Collé, whom Zinsser quotes as writing, 'It is to be hoped that this is the last air she will put on: to die in childbirth at her age, it is to call attention to herself; it is to insist on doing nothing like other people.'

Frederick on Voltaire's grief: quoted in Mercier, p 340.

Death of Émilie's baby: Zinsser, p 279.

Denis moving in with Voltaire: Vaillot, pp 319–20. V leaving for Berlin six months later: Zinsser, p 282. V falling out with Maupertuis and Frederick: Terrall, pp 302–6. Denis, the 'lady of Ferney': Soprani, p 15.

Halley's comet and the publication of Émilie's translation of the *Principia*: Zinsser, p 281.

Émilie on Halley's comet: É's commentary on the *Principia*, pp 114–15. Clairaut's prediction of the first return of Halley's comet in April 1759: see, for example, Yeomans, p 435. Newton's theory of comets: *Principia*, Book 3, for example, Propositions 40–1.

Voltaire and Jean Calas: *Chambers Biographical Dictionary*. See also Voltaire's *Traité sur la Tolérance, sur l'occasion de la Mort de Jean Calas*.

Voltaire not returning to science: Johnson and Chandrasekar, Part 2.

Voltaire's burial and funeral procession: see, for example, http://www.visitvoltaire.com.

Rousseau on politics: from his philosophical novel *Émile*, available at http://www.ilt.columbia.edu/pedagogies/rousseau/Contents2.html. On individuals submitting to the sovereign: ibid., paragraph 1655; on property: ibid., paragraph 1656; on the general will and the sovereign state: ibid., paragraphs 1648–90. Rousseau spells out these ideas further in *The Social Contract*.

Ducis on terror in the street, 'What are you talking about . . . ?': to Vallier, *Lettres du Dix-Huitième Siècle*, pp 631–2.

Émilie's grave ransacked: from Devaux's poignant recollections, summarised in Mercier, pp 341–3. See also Zinsser, p 279: there is no identification of É's grave even today.

**CHAPTER 13: Mary Fairfax Somerville**

On Maria Agnesi: I have drawn especially on Kleinert, available online at http://www.physik.uni-halle.de/Fachgruppen/history/agbas.htm, and also Findlen, 'Maria Gaetna Agnesi', in Messbarger and Findlen (eds), pp 117–19, 126–7. On Bassi: Kleinert, op. cit., and Findlen (1993).

On the growing difference between the sexes, the biology of women and exceptional women now 'unnatural': an excellent summary is provided by William E. Burns, pp 260–2. See also the references therein. In addition, see Merchant for an account of the relationship between gender and science from antiquity to the industrial and scientific revolutions of the seventeenth century.

There are two major editions of Mary Somerville's *Recollections*; the original was edited by her daughter Martha and published by John Murray in 1873. In these Notes and Sources, page numbers from this edition are denoted '*Recollections*, original edition'. But in her introduction to the 2001 '*Queen of Science*' edition of *Recollections* (listed in the Bibliography), Dorothy McMillan has given a fascinating analysis of the way Martha edited her mother's memoir in order to make her seem lovable, humble and feminine: see McMillan's introduction, pp xvi, xviii–xx. (McMillan also notes that Mary's papers were first catalogued relatively recently, by Elizabeth Chambers Patterson, who published an account of her findings in 1983: see my bibliography for full reference.) Quotes from Mary Somerville in this and subsequent chapters are from her *Recollections* (*Queen of Science* edition McMillan [ed.]), unless otherwise noted in these Notes and Sources.

Quotes from Mary's *Recollections*: Mary's love of reading, p 23; Mary on her mother, p 8; her father, pp 7 ff; Admiral Fairfax, p 58; growing up a 'wild creature', p 15; whales/dolphins, p 27; birds, pp 14, 16 (avenged by insects, p 55); helping mother: pp 15, 41; belief in witchcraft, pp 31–2, 54; 'a wild animal escaped out of a cage', p 20; detesting books on women's education 'written to please men', p 74; on British political reform in the 1790s, pp 35–6; repeal of Test Acts, p 8; cruelty and oppression in Britain, p 56. Quotes from Mary's *Recollections*, original edition: not eating sugar in protest at slavery, p 124.

On slavery: Schama, episode 12. On Wilberforce and Clarkson: *Chambers Biographical Dictionary*. On Quakers and other anti-slavery reformers: *Recollections*, p 113. The famous Enlightenment *Encyclopédie* also wrote on the slave trade: see, for example, excerpt in Tierney and Scott, Vol 2, p 114.

Locke and Habeas Corpus Act 1679: *Chambers Biographical Dictionary* (under Lord Shaftesbury).

Rousseau on women: *Émile*, paragraphs 1254–5, 1275–8; on women's virtue: ibid., paragraphs 1266–7; on men, not women, being citizens: ibid., paragraphs 1271 and translator's footnote.

Mary Wollstonecraft from *Vindication of the Rights of Woman*: on women treated as childlike, pp 81, 153; on 'gilt cage', p 131; on motherhood/roles, for example, pp 155–6; on Rousseau, pp 107 ff, 127–9 ff, 154, 189–90.

Condorcet on equal rights: quoted and translated in Lauren, p 20 (thanks to Wikipedia for alerting me to this book, which gives a global overview of the development of human rights). Condorcet's 1790 petition was presented with the help of Etta Palm d'Aelders.

Quotes from Mary's *Recollections* on her education: her father finding her a 'savage'/ school, pp 17–18; 'all a woman was expected to know', p 20; reading too much, p 23; Thomas Somerville, p 29; arithmetic, p 35; algebra, p 37; Euclid, p 38; candles/straitjacket/brother's tutor, pp 41–2; wanting to succeed, pp 46–7. For an elementary introduction to Euclid's propositions, see my *Einstein's Heroes*, pp 149–65. For an account of mathematical puzzles in women's magazines (especially *The Ladies' Diary*), see Leder.

'the Rose of Jedburgh': on p 48 of *Recollections*, Martha Somerville writes 'the Rose of Jedwood' rather than 'Jedburgh', but to avoid confusion I have used Jedburgh, since this is where Mary was born: see Patterson, p 2. Mary on her appearance: *Recollections*, pp 49, 123. Note that these passages are due to McMillan's research and were omitted from the original; for comparison, see *Recollections*, original edition, pp 61 ff, 151.

For anecdotes of Mary's teenage years (gossiping, parties, poor relations, tragedies), see *Recollections*, for example, pp 30, 32, 34, 44 (omitted from original edition), 47, 53, 58.

Mary on Jane Austen versus Mrs Radcliffe: *Recollections*, p 118 (note that I have slightly modified the last sentence).

Mary on first marriage: *Recollections*, p 60; on husband Greig: ibid., pp 63 ff (*Recollections*, original edition, pp 73–5); on being out of health after Greig's death: *Recollections*, original edition, p 77. On studying the *Principia*: ibid., p 79; on meeting Playfair: ibid., pp 81–2.

Mary on studying with John Wallace: *Recollections*, p 69 (*Recollections*, original edition, p 78). Mary on William Somerville: *Recollections*, p 73 (*Recollections*, original edition, p 87); 'extremely indignant': *Recollections*, original edition, p 88. Sermon on duties of a wife: ibid., p 88.

## CHAPTER 14: The long road to fame

Wordsworth's 'Bliss was it . . . ' is in *The Prelude XI*, lines 105–44. (I am indebted to Kramnick, p 24, for reminding me of this poem.) On Goethe and unity of British Romanticism (including Keats): I have drawn on William E. Burns, pp 116, 248. Keats's 'unweaving the rainbow' is from his poem *Lamia*, Part 2. Goethe anti-Newtonian: Boyer, p 472.

Mary on the sublime study of gravity and the heavens: *Connexion*, p 2. On the universality of the laws of physics: *Mechanism of Heavens*, Introduction. On religion: *Recollections*, pp 26–7, 115.

Mary on her scientific library: *Recollections*, p 66. On fearing her daughter's mind was strained: unpublished note quoted in McMillan, p xxxii. Martha on her mother teaching her children: *Recollections*, pp 134–5. Meeting the Herschels: *Recollections*, original edition, p 105; on the Katers, visiting Young, Wollaston calling: *Recollections*, pp 106–9 (*Recollections*, original edition, pp 128–34). For spectral lines and applications, see *Connexion*, pp 176 ff, and *Recollections*, original edition, p 218. Pioneers of emission spectra: *Connexion*, p 312; cause of spectra unknown: ibid., p 178; modern explanation: Mitchell (ed.), p 80, Herb Fried, private communication.

Wollstonecraft and Rousseauist wives: Schama, episode 12. De Staël: see Byron Study Centre website at http://byron.nottingham.ac.uk/.

Seeing Voltaire's statue, and Mary's letter on French ladies: quoted in Patterson, p 22.

On Sophie Germain (biographical overview, including Fourcroy's address to the National Convention, her Academy essays, Biot's 1817 review): Stupuy's preface to Sophie's *Oeuvres* (1896). Sophie first woman to attend Academy meetings: Fara, p 96. Sophie to Gauss, fear of *femme savante* label, etc.: *Cinq lettres*... Gauss to Sophie on discovering her identity: excerpted and translated in Fauvel and Gray (eds), pp 497–8. For a mathematical overview of Sophie's work on Fermat's 'last theorem', see Riddle's article at http://www.agnesscott.edu/lriddle/women/germain-FLT/SGandFLT.htm. For a detailed historical analysis of her work, see Laubenbacher and Pengelley's article at http://en.scientificcommons.org/55651200.

On the Committee on Weights and Measures: Boyer, pp 471–2; on standard metre bar: Resnick and Halliday, p 5.

On Lagrange's response to Lavoisier's execution: quoted and translated in Bell, p 118. On Lavoisier: for an interesting article, see 'Antoine Laurent Lavoisier' at http://cti.itc.virginia.edu/~meg3c/classes/tcc313/200Rprojs/lavoisier2/home.html.

On Marie Lavoisier's scientific contribution: Rebière, p 60; on her imprisonment: Fara, pp 86–91.

Mary on fear of Napoleonic wars/bad harvests: *Recollections*, p 59 (*Recollections*, original edition, p 71). History of Napoleon's defeat, restoration of monarchy, republic, etc.: *Oxford Reference Encyclopedia.*

Mary on dinner at Laplace's, and meeting the marquise in bed: *Recollections*, pp 89–90; Brougham's letter and Mary's response: ibid., pp 131–2; Mary on domestic interruptions: ibid., p 133; Laplace's letter on Newton: ibid., p 148; Mary on Babbage: ibid., p 115; Mary on Ada Lovelace: ibid., pp 125–6. On Babbage's machine and Ada Lovelace: Perl, pp 101–25. On Mary and Ada: Patterson, pp 148–50.

British calculus reform, Newton's shadow and notation not to blame: Boyer, pp 459–60.

Bowditch on Laplace: *Memoir*, p 62. (*Memoir*, by Bowditch's children, was published with the first volume of Bowditch's translation of *Celestial Mechanics*.) Bowditch on Mary Somerville: ibid., pp 68, 167. Mary on Bowditch: *Recollections*, p 180.

On John Murray: Patterson, p 74, and *Chambers Biographical Dictionary*. John Murray and De Staël: *Chambers Biographical Dictionary*, and *The Literary Encyclopedia* at http://www.litencyc.com/. Mary on John Murray: *Recollections*, p 176.

Mary on outcry over geology of Darwinism: *Recollections*, p 106 (*Recollections*, original edition, p 129). Mary on Darwin not explaining 'primordial forms': ibid., p 288. This passage, which also contained Mary's summary of and support for the general theory of evolution, was cut from the original.

Responses to Mary's book, and her reaction: Patterson, pp 83 ff, including Poisson, p 87; Lady Herschel and Mary Kater on Royal Society bust: ibid., p 90. Mary's mother's supportive response: McMillan, p xxxii. Peacock, Whewell, Somerville and Biot's responses: *Recollections*, pp 139–40, 142, 143 (*Recollections*, original edition, pp 172–5). *Mechanism of the Heavens* a standard text for one hundred years: Rebière, p 61, and Fara, p 108.

On women's colleges: see the websites of Girton and Somerville colleges.

**CHAPTER 15:** ***Mechanism of the Heavens***

Maria Edgeworth to Mary: *Recollections*, original edition, pp 204–5.

On calculus of variations: Mary's elementary account was concise and clear; Bowditch gave a more extensive account, but he, too, used elementary notions of Leibnizian calculus, rather than the Euler-Lagrange equations. Note that Maupertuis's colleague Euler eliminated the metaphysical from his development of the calculus of variations. For a modern treatment of the topic, including the way Bernouilli adapted Fermat's principle to find the brachistochrone, see Weinstock, pp 19, 67–71.

Principle of least action: today, this 'active' quality is often related to kinetic energy, but this concept was not yet properly developed in Mary's day; she and Laplace used Euler's integral version of Maupertuis's definition of 'action' (namely, velocity multiplied by distance). For a modern 'metaphysical' view of least action (towards an end rather than a cause), see Magueijo, pp 186–7.

Laplace on God as hypothesis: Boyer, p 494. Laplace's use of principle of least action: *Celestial Mechanics*, for example, pp 139, 172, 265.

Laplace's version of the 'principle of living forces': *Celestial Mechanics*, p 99.

Émilie to Maupertuis on *vis viva*: Besterman, letter numbers 122, 124, April to May 1738. Note that É was initially worried that conservation of 'living force' would impose a limitation on free will, because it might restrict the amount of motion we are free to undertake. Émilie on *vis viva*: *Institutions*, for example, pp 430–2.

Leibniz on *vis viva*, conservation/raising weights: extracted in Wiener (ed.), pp 133–7, 182, 314–16. Leibniz on Galileo's contribution: ibid., p 127.

Émilie on Newton (supposedly) disagreeing with living force: *Institutions*, pp 444–5.

Newton on 'work' and conservation of energy: *Principia*, Propositions 39–41, plus cryptic paragraph on 'rate of work' and machines: scholium to Newton's laws, p 430. Thomson and Tait's response to this significant paragraph, and also Maxwell's response: Chandrasekhar, pp 34–5. For more on Tait's interpretation (from a later work on thermodynamics, published 1877), see Cohen (1999), p 119. 'Did Newton have the concept of energy?': Cohen (1999), pp 119–22, and Chandrasekhar, Chapter 9.

Young on the terms 'labour' and 'energy': Young, Lecture VII, pp 59–60. (I was alerted to this passage by Patterson, p 124.) Note that Gaspard Gustave de Coriolis later introduced the term 'work' in France, in 1829: see, for example, O'Connor and Robertson's biography of Coriolis at http://www-history.mcs.st-andrews.ac.uk/Biographies/Coriolis.html; see also Grattan-Guiness, p 165.

Thomson (and Tait) popularising concept of conservation of energy: Maxwell, Vol 2, p 200. Thomson and Carnot: Resnick and Halliday, p 629.

Leibniz on non-elastic collisions and intuitive statement of 'universal conservation of energy': Leibniz, *New Essays Concerning Human Understanding*, Appendix ('Essay on Dynamics . . . in which it is shown . . . absolute force [is preserved] . . . '), p 670. (Note Leibniz confusing mass and weight: ibid., p 659.) Others on conservation of $mv^2$: Huygens had shown this was true for 'elastic' collisions, and Wallis and Wren each independently came up with similar preliminary results at the same time, but Leibniz generalised the idea, albeit only intuitively, to conservation of total energy. See Leibniz, extracted in Wiener (ed.), p 127. See Huygens, *On the motion of bodies resulting from impact*, translated by Michael S. Mahoney, available online.

Joule's experiment, and heat as energy: my description is based on the information in Resnick and Halliday, pp 554–6. Experimental verification of conservation of total energy: ibid., p 547. Relativistic mass-energy: nature of internal energy: Resnick, p 136. Sun's conversion of mass to light (two kilograms per second): Taylor and Wheeler, p 244. Everyday gain in mass through heat too small to detect: Resnick, p 138, and Taylor and Wheeler, p 223 (note this reference describes both Thompson's negative experiment and a potential method that might enable such tiny mass gains to be measured).

'Mass' or 'rest mass': generally the mass associated with 'rest energy' is called 'rest mass', but choosing language to convey the meaning of Einstein's equation by analogy with Newtonian concepts is conceptually fraught: see, for example, D'Inverno, p 45, and Taylor and Wheeler (T&W), p 251. T&W say 'rest energy' equals 'mass' (p 201, equation 7.13; p 193) to reflect the Newtonian idea of mass: they say a photon's mass is zero, but it can *transfer* mass to a material object (pp 232, 246 and p 226 for energy of *motion*). On the other hand, D'Inverno says 'rest energy' equals 'rest mass', that is, in relativistic units where $c = 1$ (pp 45, 48), but that a relatively moving photon's 'relativistic mass' is not zero (p 50). To avoid confusion for lay readers, I have followed T&W (p 251, 'mass is mass is mass!'). All agree, however, that energy and mass are only 'equivalent' in the particle's rest frame. It is a matter of language, not mathematics: for example, D'Inverno's 'relativistic mass' (p 46) is T&W's 'relativistic energy' (p 201, equation 7.11). 'Intrinsic' mass is another name for 'rest mass': Penrose, p 282.

Newton's 'Query 31' in *Opticks*: solar system needing re-forming because of perturbations, p 402; hard atoms, 'conservation of mass', p 400; motion not conserved, p 398, so need to 'recruit' it so world doesn't grow cold (that is, intuitive 'conservation of energy', with apparently alchemical influences), pp 399–400; non-elastic collisions, p 398.

Leibniz's letters to Samuel Clarke: extracted in Wiener (ed.), especially p 216 (winding up watch).

Conservation of mechanical energy for a single orbiting planet: Resnick and Halliday, p 412.

Mary on stability: *Mechanism of the Heavens*, preliminary dissertation, pp xv ff, and *Connexion*, pp 15–29. The first sentence in Mary's passage that I have quoted on pp208–9 (from *Connexion* p28) is not quite correct (eg resonance occurs *near* conjunction not *at* it); my thanks to Rosemary Mardling for clarifying my section on this topic.

General Relativity (GR) and the perihelion of Mercury: the GR prediction is accurate to within experimental error: D'Inverno, p 198, table 15.1. Further mathematical references include Fowles, p 138. Einstein's ecstasy: quoted in Hoffmann and Dukas, p 125. The range of applicability of the inverse-square law is still being debated; for instance, observed deviations from the Newtonian-Einsteinian models are currently being explained by assuming the existence of 'dark matter', or of extra, hidden dimensions in spacetime. None of these 'theories' is complete, so, at this stage, GR is still the definitive theory of gravity, although it needs modifying because it breaks down at 'singularities'.

On chaos theory in the solar system: see Peterson for popular overview. For more recent technical details: Batygin and Laughlin; Laskar and Gastineau; Tremaine's online lecture on stability at http://www.astro.princeton.edu/~tremaine/alex/lecture1.ppt; Sansottera et al. at http://arxiv.org/PS_cache/arxiv/pdf/1010/1010.2609v1.pdf). On chaos theory in general, including snowflakes: Gleick.

**CHAPTER 16: Mary's second book: popular science in the nineteenth century**

Mary's children's letters on her reception in Paris: quoted in Patterson, pp 95, 107. Meeting Henry Bowditch: ibid., p 110. Woronzow as FRS: McMillan, p 336.

Mary on the Lafayettes: *Recollections*, p 150 (*Recollections*, original edition, p 184). On the genesis of her *Connexion*: *Recollections*, p 146. On Madame Rumford: ibid., p 152 (*Recollections*, original edition, p 188). On De Staël's daughter: *Recollections*, original edition, p 187.

Maxwell on *Connexion*: quoted in Patterson, p 98.

Selections from *Conversations on Chemistry* (1817 edition): for example, reflection and absorption, pp 67 ff; conduction, pp 71 ff; steam, pp 80–121 ff.

*Conversations on Chemistry* today: tin box described in John H. Lienhard, episode 744: 'Engines of Ingenuity', University of Houston, at http://www.uh.edu/engines/epi744.htm. Lienhard presented this as an example he would use in his own classes. In episode 1302, Lienhard discussed Marcet's son, Arago, and the steam pressure device. Also, Rossotti has recently published a book of extracts from Marcet called *Chemistry in the Schoolroom* (AuthorHouse, Bloomington, 2006).

Mary on Faraday and Marcet: *Recollections*, p 92.

*Connexion* a best seller: number of editions/translations: *Recollections*, original edition, pp 202–3. For more details, see Patterson, pp 146, 177. Mary's royalties: ibid., pp 136, 146, 192 (and fifteen thousand copies sold all up, p 193). Murray commissioning Mary's portrait: ibid., p 142. Sales figures on Jane Marcet: Rossotti, p 61.

Mary signing Martineau's copyright petition: Patterson, p 141. Other petitions: *Recollections*, p 279.

Mary's pension: *Recollections*, pp 144–5. For the political implications of the award, see Patterson, Chapter 8. Woronzow's letter: quoted in Patterson, p 163; Mary's wry comment on pension: quoted in ibid., p 179; Mary on 'tame Lioness': quoted in ibid., p 12. Mary on Queen Victoria: *Recollections*, original edition, pp 147, 203. Mary actually said Victoria was eighteen years old at coronation, which was the age at which she had become queen, but the coronation ceremony itself took place just after she turned nineteen.

Jane Marcet on Mary's modesty: *Recollections*, original edition, p 210. Smyth on Halley's comet/Newton: ibid., p 211. Return of Halley's comet in 1835: ibid., p 100. Gauss predicting orbit of asteroid: Peterson, pp 175–8; Piazzi: ibid., pp 175, 177. Note that W. Herschel provided the name 'asteroid': ibid., p 177.

Mary on weighing the sun: *Connexion*, note 134.

Faraday's corrections of *Connexion*: *The Correspondence of Michael Faraday* (ed. James), Vol 2, letter numbers 684, 732.

Herschel on Mary's first experiment (light and magnetism): quoted in Patterson, p 47; Mary's mortification on experiment disproved: ibid., p 48. Maxwell's assessment of Mary's experiment: Maxwell, Vol 2, p 451. Voltaire linking light and magnetism: Bodanis, p 128.

Mary's experiment on colour and chemistry: for a historical overview, see Patterson, pp 173–4. Mary on Melloni: *Connexion*, pp 235–40; Mary on 'chemical' or 'photographic' rays: *Connexion*, pp 224–33. Faraday and Mary's correspondence on the experiment: *The Correspondence of Michael Faraday* (ed. James), Vol 2, letter numbers 821, 824.

Herschel on Mary's third experiment (colour and juices): *Recollections*, pp 226–7 (*Recollections*, original edition, pp 278–9).

Whewell on Mary 'a person of real science': quoted in Patterson, p 138. Faraday thanking Mary for her insight: *The Correspondence of Michael Faraday* (ed. James), Vol 2, letter number 701. Mary's book cutting-edge: as another example, Patterson notes that in the second edition of *Connexion*, Mary gave what was probably the very first popular account of Faraday's discovery and analysis of electrolysis: Patterson, p 134.

**CHAPTER 17: Finding light waves: the 'Newtonian revolution' comes of age**

Mary on light: *Connexion*, pp 172–218. No annihilation of particles: ibid., pp 183–4. More on Young's experiment: Halliday and Resnick, pp 1148, 1068 ff.

History of diffraction: *Connexion*, pp 189–90; interference pattern in diffraction: ibid., note 197. See also Resnick and Halliday, p 1099, and for diffraction through pinhole, p 1112; interference (technically 'superposition') in diffraction pattern: ibid., pp 1118, 1119–20. The effects of diffraction had first been discussed by Francesco Grimaldi in the seventeenth century.

Newton's analysis of sound: *Principia*, Book 2. Westfall, p 455, calls it a 'minor triumph'.

Mary's description of dependence of interference on wavelength: *Connexion*, p 184. Newton's genius in measuring: ibid., p 189.

Young's detractors: Resnick and Halliday, p 1074; Foucault on speed of light: ibid., p 1023.

Laplace on black holes: an English translation of Laplace's essay on the conditions for light to be unable to escape the force of gravity of a star is reprinted in Hawking and Ellis, pp 365–8. For a summary, see D'Inverno, p 224. On John Michell: Davies, pp 108–9, 116. Note that in 1916 Karl Schwarzschild found the first relativistic black hole 'solution' of Einstein's equations: he solved this equation while he was in the trenches fighting for Germany in the First World War, in which he was later killed. At present, evidence for the existence of black holes is indirect. But see *New Scientist*, 18 September 2010, for new search.

Newton's wave-particle duality: *Opticks*, for example, Part 2, Proposition XII; Newton on light and matter interconvertible: *Opticks*, 'Query 30'. For commentary, see Cohen's preface to the 1952 Dover Publications edition of *Opticks*, pp xlvii, p xl, and the introduction by Whittaker, p lxiii. (This edition also has a tribute to Newton by Einstein.) Newton directly inspiring Young, as much as Huygens: Cohen's preface to *Opticks*, pp xl, xlviii–xlix. Young using Newton's data: ibid., p xli. On Newton's confusion, for example, ether or not: Westfall, pp 522–3.

On wave-particle duality and the two-slit experiment: interpreting quantum mechanical experiments is problematic, so for excellent, slightly different, relatively introductory interpretations of the two-slit results, see Penrose, pp 300–4, and Davies and Gribbin, pp 201 ff. I have drawn on both these references in my brief summary, and also Resnick and Halliday, pp 1200 ff. Note that these dual effects are 'real': for example, electron 'waves' are harnessed in electron microscopes: Davies and Gribbin, p 200.

Mary on ether proved: *Connexion*, pp 24, 184. Note that she suggested ether may also ultimately affect planetary orbits: ibid., pp 191–4. Mary on electricity and ether: ibid., p 304.

Electrostatic charge (glass and silk): Mitchell (ed.), p 55. Franklin's labels: Resnick and Halliday, p 648. Volta's battery: note that an ancient device that could have functioned as a weak battery was found in Baghdad.

Mary on electromagnetism: *Connexion*, pp 350 ff; on Ampère: pp 356 ff. Faraday on the connection between heat and electricity: ibid., p 333; on unity of forces: ibid., p 377; on speed of electricity: ibid., p 317; on inverse-square law: ibid., p 308; electric vibrations in ether: ibid., p 304; earth's magnetism: ibid., pp 370–1; on the rotary action of electromagnetic forces: ibid., pp 351–2 (and straight line forces and composition of forces: ibid., pp 10, 351); on the galvanometer: ibid., p 354; on Faraday's prototype motor: ibid., pp 351–2; on Faraday's coil: ibid., p 361; on Galvani and Volta: ibid., pp 324–5; Henry's electromagnet: ibid., p 354.

Joseph Henry's response to Mary and her *Connexions*: quoted in Patterson, pp 182–3; see also Hamilton, p 275.

Mary on Coulomb: *Connexion*, p 308. Priestley's analogy: Halliday and Resnick, pp 693–4.

More on Ampère's contribution: Maxwell, Vol 2, Article 502 ff. See also O'Connor and Robertson's biography of Ampère at http://www-groups.dcs.st-and.ac.uk/~history/Biographies/Ampere.html; this article also mentions that Laplace and Biot favoured corpuscular theory of light.

On the development of Faraday's field idea and Maxwell's electromagnetic theory: I have given a much more detailed popular account in my *Einstein's Heroes*. For Maxwell's own account of his (truly Newtonian) method, see his *Treatise on Electricity and Magnetism*, Vol 2, Articles 522, 592; on action-at-a-distance versus field maths: ibid., pp 176–7, 198 ff; on Maxwell's use of the principle of least action (or more specifically, of the Lagrange-Hamilton equations consistent with the least action principle): ibid., p 198, plus Chapters V–VII and beyond.

## CHAPTER 18: Mary Somerville: a fortunate life

Woronzow's secret child: McMillan, p xxxiv. Girls' sailing boat: *Recollections*, original edition, p 337.

Mary on geological age of earth, York cathedral, etc.: *Recollections*, p 106 (*Recollections*, original edition, p 129). Medals for *Physical Geography*: *Recollections*, original edition, pp 350–1. Mary repenting *Molecular Science*: ibid., p 338.

J.S. Mill to Mary: *Recollections*, original edition, p 278. Mary on Mill (and US insult to women/petition): ibid., p 277 (*Recollections*, original edition, pp 344–6). Women as slaves: see Mill's *On the Subjection of Women*, especially Chapters I and 2; on violence, property: ibid., Chapter 2. Property Act 1882: extracted in Walsh, pp 350–51. Schama, episode 13, mentions that on his marriage to Harriet, Mill publicly renounced his own property rights. Schama also mentions that all British male householders had the vote since 1867.

Mary's last years, in *Recollections*, original edition: Darwin/birds/treatment of wives: pp 357–9. Laplace's manuscripts thrown in river: p 360. Professor Pierce, studying higher mathematics: p 356; novels (not tragic): p 357; memory for maths still good: p 364; revising old manuscript: p 202; pet bird: pp 332, 352; Frances Cobbe on William's death: p 326; regretting the sky/scientific expeditions: pp 343–4, 348; slavery/blue Peter: pp 373–4; Martha, 'joyous spring': p 376; Emma Chenu and 'Russian lady': pp 345–6. Garibaldi: p 336 (*Recollections*, pp 245, 271). (Frere and anti-slavery treaty: *Chambers Biographical Dictionary*.)

**EPILOGUE: Declaring a point of view**

Gender statistics: see Appendix

Accuracy of general relativity: Penrose, p 198, and D'Inverno, p 192.

# BIBLIOGRAPHY

Most of Émilie du Châtelet's works and many of her letters are available online from the Bibliothèque Nationale de France (BnF) Gallica library. I have also greatly benefited from the copy of Émilie's *Institutions de physique* at the Women's Online Network, Miami University of Ohio. Web addresses of other online references listed below are given in my Notes and Sources.


M. Algarotti, *Le Newtonianisme pour les dames, ou Entretiens sur la lumière*, translated from Italian by Duperron de Castera, Chez Montalant, Montpellier, 1738.

Robyn Arianrhod, *Einstein's Heroes*, University of Queensland Press, St Lucia, 2003; Oxford University Press, New York, 2005.

Elisabeth Badinter, *Émilie, Émilie: L'Ambition feminine au XVIIIe siècle*, Flammarion, Paris, 1983.

K. Batygin and G. Laughlin, 'On the dynamical stability of the solar system', *Atrophysical Journal*, Vol 683, No 2, 2008, pp 1207 ff.

Eric Temple Bell, *Mathematics: Queen and Servant of Science*, Tempus Books, Washington, 1987 (originally published 1951).

Barbara M. Benedict, 'The mad scientist: The creation of a literary stereo-type', in Robert C. Leitz III and Kevin L. Cope (eds), *Imagining the Sciences: Expressions of New Knowledge in the 'Long' Eighteenth Century*, AMS Press, New York, 2004.

David Bodanis, *Passionate Minds: The Great Enlightenment Love Affair*, Little, Brown, New York, 2006.

Carl B. Boyer, *The History of the Calculus and its Conceptual Development* (*The Concepts of the Calculus*), Dover Publications, New York, 1959. In my Notes and Sources, this reference is denoted by 'Boyer (1959)'.

Carl B. Boyer (revised by Uta C. Merzbach), *A History of Mathematics*, John Wiley & Sons, New York, 1991. In my Notes and Sources, this reference is denoted by 'Boyer'.

# *Bibliography*


William E. Burns, *Science in the Enlightenment: An Encyclopedia*, ABC-CLIO, 2003.

Steven M. Cahn (ed.), *Classics of Western Philosophy*, Hackett Publishing Co., Indianapolis, 1990. (This is an excellent collection of original philosophical works, in English.)

S. Chandrasekhar, *Newton's Principia for the Common Reader*, Clarendon Press, Oxford, 1995. In my Notes and Sources, this reference is denoted by 'Chandrasekhar', as distinct from 'Johnson and Chandrasekar' (see below, different spelling).

Émilie du Châtelet, commentary on Newton's *Principia*, Book 2 of her translation: *Principes Mathématiques de la Philosophie Naturelle* par feue Madame la Marquise du Chastelet, Desaint & Saillant, Lambert, Paris, 1759; facsimile reproduced in 1990 edition by Éditions Jacques Gabay.

—, correspondence: *Lettres de Voltaire et de sa célèbre amie [la mise Du Châtelet]; suivies par . . . Rousseau*, Cailleau (Genève), Geneva, 1782. (This collection includes letters from both Voltaire and Émilie.) In my Notes and Sources, this reference is denoted by *Lettres de Voltaire et de sa célèbre amie*

—, correspondence: *Lettres de la marquise du Châtelet, du roi de Prusse, et de Voltaire (or 'Lettres inédites de Mme la M'ise du Châtelet, et supplement à la correspondance de Voltaire avec le Roi de Prusse et avec différentes personnes célèbres')*, Lefebvre, Paris, 1818. BnF Document Number Z 47358. In my Notes and Sources, this reference is denoted by *Lettres Inédites*.

—, correspondence: *Lettres inédites de Mme la M'ise du Chastelet à M. le comte d'Argental; auxquelles on a joint une Dissertation sur l'existence de Dieu; Les réflexions sur le bonheur / par le même auteur, et deux notices historiques sur Mme Du Chastelet et M. d'Argental / [par Hochet]*, Chrouet, Paris, 1806. In my Notes and Sources, this reference is denoted by 'BnF Document Number Z 15192'.

—, correspondence: *Les Lettres de la Marquise du Châtelet*, Theodore Besterman (ed.), Institut et Musée Voltaire, Les Délices, Geneva, 1958. In my Notes and Sources, this reference is denoted by 'Besterman'.

—, correspondence: *Lettres d'Amour au Marquis de Saint-Lambert*, Anne Soprani (ed.), Collection Cachet Volant, Éditions Paris-Méditerranée, Paris, 1997. In my Notes and Sources, this reference is denoted by *Lettres d'Amour*.

—, correspondence: *Quelques lettres inédites de la M'ise du Châtelet et la duchesse de Choiseul*, Ernest Jovy, H. Leclerc, Paris, 1906.

—, *Discours sur le bonheur*, preface by Elisabeth Badinter, Éditions Payot et Rivages, Paris, 1997.

—, *Discours sur le bonheur*, edited and with a critical commentary by Robert Mauzi, Société d'editions 'Les Belles Lettres', Paris, 1961.

—, *Dissertation sur la nature et la propagation du feu*, Chez Prault Fils, Paris, 1744. (This is the expanded later edition of Émilie's essay.)

—, preface and translation, *La fable des abeilles* (Émilie's French translation of part of Bernard Mandeville's *Fable of the Bees: or, Private Vices, Publick Benefits*, J. Roberts, London, 1714), in Wade, *Studies on Voltaire, with Some Unpublished Papers of Mme du Châtelet*, Princeton University Press, Princeton, 1947.

—, *Institutions de physique*, Chez Prault Fils, Paris, 1740. This is referred to as *Fundamentals of Physics* in the narrative.

—, *Réponse de madame la marquise du Châtelet à la lettre que m. de Mairan lui a écrite le 18 février 1741 sur la question des forces vives*, originally published by Foppen, Bruxelles, in 1741, then republished with her essay on fire by Chez Prault Fils, Paris, 1744. (This is Émilie's famous response to Mairan on the question of *vis viva*.)

I. Bernard Cohen, 'A Guide to Newton's *Principia*', published at the beginning of *Isaac Newton: The Principia: Mathematical Principles of Natural Philosophy*, A New Translation by I. Bernard Cohen and Anne Whitman, assisted by Julia Budenz, University of California Press, Berkeley, 1999. In my Notes and Sources, my references to Cohen's guide – as opposed to Newton's *Principia* – are denoted by Cohen (1999).

—, *Introduction to Newton's 'Principia'*, Cambridge University Press, Cambridge, 1971.

Kevin L. Cope, 'Elastic empiricism, interplanetary excursions and the description of the unseen', in Robert C. Leitz III and Kevin L. Cope (eds), *Imagining the Sciences: Expressions of New Knowledge in the 'Long' Eighteenth Century*, AMS Press, New York, 2004.

Paul Davies, *About Time*, Viking, London, 1995.

— and John Gribbin, *The Matter Myth: Towards 21st Century Science*, Viking, London, 1991.

Madame du Deffand, *Lettres à Voltaire*, Rivages Poche, Paris, 1994.

Michel Delon, 'Voltaire irrité et Diderot séduit', in *Magazine Littéraire*, No 416, January 2003.

René Descartes, *Meditations on First Philosophy*, in Steven M. Cahn (ed.), *Classics of Western Philosophy*, Hackett Publishing Co., Indianapolis, 1990, pp 405–45.

Russell R. Dynes, *The Dialogue Between Voltaire and Rousseau on the Lisbon Earthquake: The Emergence of a Social Science View*, Disaster Research Center, Department of Sociology and Criminal Justice, University of Delaware, 1999.

Esther Ehrman, *Madame du Châtelet: Scientist, Philosopher and Feminist of the Enlightenment*, Berg, Oxford, 1986.

Jean-Luc Faivre, *Voltaire je connais!*, Mallard Editions, Paris, 1998.

Patricia Fara, *Scientists Anonymous: Great Stories of Women in Science*, Wizard Books, Cambridge, 2007.

J. Fauvel and J. Gray (eds), *The History of Mathematics: A Reader*, Open University/ Macmillan Education, 1987. (This has an excellent variety of excerpts in English from original sources.)

Paula Findlen, 'Science as a career in Enlightenment Italy: The strategies of Laura Bassi', in *Isis: An International Review Devoted to the History of Science and Its Cultural Influences*, Vol 84, No 3, 1993.

—, 'Becoming a scientist: gender and knowledge in eighteenth-century Italy', in *Science in Context*, Vol 16, 2003, pp 59–87.

Grant R. Fowles, *Analytical Mechanics*, Holt, Rinehart and Winston, New York, 1966.

Elizabeth Garber, *The Language Of Physics: The Calculus and the Development of Theoretical Physics in Europe, 1750–1914*, Birkhaüser, Boston, 1999.

Sophie Germain, *Cinq Lettres de Sophie Germain à Carl Frederic Gauss*, 1804–07, microfilm, BnF.

—, *Considérations Générales sur L'état des Sciences et des Lettres*, microfiche, BnF.

—, *Oeuvres philosophiques de Sophie Germain; suivies de pensées et de lettres inédites et précédées d'une notice sur sa vie et ses oeuvres par Hte Stupuy*, Firmin-Didot, Paris, 1896.

James Gleick, *Chaos*, Penguin, New York, 1987.

Olympe de Gouges, *Declaration of the Rights of Woman and the Female Citizen*, 1791, quoted and translated in Tierney and Scott, pp 190–2.

Françoise de Graffigny, *Correspondance de Madame de Graffigny*, J.A. Dainard, et al. (eds), 10 Vols, Voltaire Foundation, Oxford, beginning 1985.

I. Grattan-Guinness, 'Does History of Science treat of the history of science? The case of mathematics', *History of Science*, Vol 28, 1990, pp 149–73.

Alfred Rupert Hall, *Philosophers at War: The Quarrel between Newton and Leibniz*, Cambridge University Press, Cambridge, 2002.

Frank Hamel, *An Eighteenth Century Marquise: A Study of Émilie du Châtelet and her Times*, Stanley Paul and Co., London, 1910.

James Hamilton, *Faraday: The Life*, Harper Collins, London, 2002.

S.W. Hawking and G.F.R. Ellis, *The Large-Scale Structure of Spacetime*, Cambridge University Press, Cambridge, 1973.

Miguel de Icaza Herrera, 'Galileo, Bernouilli, Leibniz and Newton around the brachistochrone problem', in *Revista Mexicana de Física*, Vol 40, No 3, 1994, pp 459–75.

Banesh Hoffmann with Helen Dukas, *Einstein*, Paladin, Frogmore, 1977.

*How Is It Done?* Reader's Digest Association Ltd, London, 1995.

Ray d'Inverno, *Introducing Einstein's Relativity*, Clarendon Press, Oxford, 1993.

Frank James (ed.), *The Correspondence of Michael Faraday*, Vol 2, 1832–40, Institution of Electrical Engineers, London, 1993.

W. Johnson and S. Chandrasekar, 'Voltaire's contribution to the spread of Newtonianism – I. Letters from England: Les Lettres Philosophiques', *International Journal of Mechanical Science*, Vol 32, No 5, 1990, pp 423–53.

—, 'Voltaire's contribution to the spread of Newtonianism – II. Éléments de la philosophie de Neuton: The Elements of the Philosophy of Sir Isaac Newton', in *International Journal of Mechanical Science*, Vol 32, No 6, 1990, pp 521–46.

Immanuel Kant, *Prolegomena to Any Future Metaphysics*, in Steven M. Cahn (ed.), *Classics of Western Philosophy*, Hackett Publishing Co., Indianapolis, 1990, pp 933–1008. (This is a later (1783), shorter version of ideas in his *Critique of Pure Reason.*)

Andreas Kleinert, 'Maria Gaetana Agnesi und Laura Bassi: Zwei italienische gelehrte Frauen im 18. Jahrhundert', in *Frauen in den exakten Naturwissenschaften*, Willi Schmidt and Christoph J. Scriba (eds), Steiner, Stuttgart, 1990, pp 71–85. (Available online from University of Halle.)

Morris Kline, *Mathematics: The Loss of Certainty*, Oxford University Press, New York, 1980.

Miriam Brody Kramnick, introduction to *Vindication of the Rights of Woman*, Mary Wollstonecraft, Penguin Classics, international edition, 1978.

Huguette Krief, 'Marie du Deffand, Isabelle de Charrière', in *Dix-Huitième Siècle (Femmes des Lumières)*, No 36, 2004, p 282.

Rotraud von Kulessa, 'Vertu et Sensibilité', in *Dix-Huitième Siècle (Femmes des Lumières)*, No 36, 2004, pp 211 ff.

Pierre-Simon Laplace, *Celestial Mechanics*, annotated and translated into English by Nathaniel Bowditch, Chelsea Publishing Co., New York, 1966.

J. Laskar and M. Gestineau, 'Existence of collisional trajectories of Mercury with Mars and Venus with the Earth', *Nature*, No 459, 2009, pp 817–19.

Reinhard Laubenbacher and David Pengelley, '"Voici ce que j'ai trouvé": Sophie Germain's grand plan to prove Fermat's Last Theorem', *Historia Mathematica*, 2010.

Paul Gordon Lauren, *The Evolution of International Human Rights*, University of Pennsylvania Press, Pennsylvania, 2003.

Gilah C. Leder, 'A diary fit for every lady's toilet and gentleman's pocket', in *Function: A School Mathematics Magazine*, Monash University, 1985, pp 7–12.

Gottfried Leibniz, *New Essays Concerning Human Understanding*, translated and annotated by Alfred Gideon Langley, The Open Court Publishing Company, Illinois, 1949. This includes Leibniz's *Essay on Dynamics* as an appendix. (Other Leibniz papers – including *Discourse on Metaphysics*, *The Mondadology*, *The Theodicy*, *Essay on Dynamics*, and letters to Samuel Clarke – are extracted in Wiener (ed.), see below.)

Robert C. Leitz III and Kevin L. Cope (eds), *Imagining the Sciences: Expressions of New Knowledge in the 'Long' Eighteenth Century*, AMS Press, New York, 2004.

John Locke, *An Essay Concerning Human Understanding*, abridged version, in Steven M. Cahn (ed.), *Classics of Western Philosophy*, Hackett Publishing Co., Indianapolis, 1990, pp 617–712.

Koffi Maglo, 'The reception of Newton's gravitational theory by Huygens, Varignon, and Maupertuis: How normal science may be revolutionary', *Perspectives on Science*, Vol 11, No 2, pp 135–69, Massachusetts Institute of Technology, 2003.

Dorothy McMillan (ed.), *Queen of Science: Personal Recollections of Mary Somerville*, Canongate, Edinburgh, 2001. In my Notes and Sources, McMillan's introduction to this work is denoted by 'McMillan', while Mary Somerville's reminiscences are denoted by '*Recollections*' (referring to McMillan's expanded *Queen of Science* edition), and/or '*Recollections*, original edition', see Somerville below.

Magnus Magnusson (ed.), *Chambers Biographical Dictionary*, Chambers, Edinburgh, 1990.

João Magueijo, *Faster than the Speed of Light*, Perseus, Cambridge, 2003.

Jean-Jacques d'Ortous de Mairan, *Lettre à Mme*** sur la question des forces vives [en réponse]*, Académie Royale des Sciences, Paris, 1741.

Jane Marcet, *Conversations on Chemistry*, 5th edition, Longman, Hurst, Rees, Orme and Brown, London, 1817.

Pierre Louis Moreau de Maupertuis: *Discours sur les différentes figures des astres*, De l'Imprimerie Royale, Paris, 1732. It was published in English in 1734, in an appendix to a book called *Dr Burnet's Theory of the Earth* . . . by J. Keill, Oxford, as 'A Dissertation on the Different Figures of the Celestial Bodies, and from thence, some conjectures concerning the stars which seem to alter their magnitude; and concerning Saturn's ring. With a summary exposition of the Cartesian and Newtonian systems by Monsieur Maupertuis'.

James Clerk Maxwell, *Treatise on Electricity and Magnetism*, Vols 1 and 2, Dover Publications, New York, 1954 (reprint of 3rd edition published by Clarendon Press, Oxford, 1891).

Massimo Mazzotti, *Newton for Ladies*, Bologna Science Classics Online, 2003.

Carolyn Merchant, *The Death of Nature: Women, Ecology and the Scientific Revolution*, Harper & Row, San Francisco, 1983.

Gilbert Mercier, *Madame Voltaire*, Éditions de Fallois, Paris, 2001.

Rebecca Messbarger and Paula Findlen (eds), *The Contest for Knowledge: Debates over Women's Learning in Eighteenth Century Italy*, University of Chicago Press, Chicago, 2005.

Mary Midgley, *Science as Salvation: A Modern Myth and its Meaning*, Routledge, London, 1992.

Charles W. Misner, Kip S. Thorne and John Archibald Wheeler, *Gravitation*, W.H. Freeman & Co., San Francisco, 1973.

James Mitchell (ed.), *The Illustrated Reference Book of Science*, Colporteur Press, Sydney, 1982.

Paul Veatch Moriarty, 'The principle of sufficient reason in Du Châtelet's Institutions', in *Studies on Voltaire and the Eighteenth Century*, No 1, 2006.

Isaac Newton, *The Principia: Mathematical Principles of Natural Philosophy*, translated by I. Bernard Cohen and Anne Whitman, assisted by Julia Budenz, University of California Press, Berkeley, 1999.

—, *Opticks*, Dover Publications, New York, 1952, with foreword by Albert Einstein, introduction by Sir Edmund Whittaker and preface by I. Bernard Cohen (based on the 4th edition, London, 1730).

J.J. O'Connor and E.F. Robertson, MacTutor History of Mathematics Archive, School of Mathematics and Statistics, University of St Andrews, online. (This includes online biographies of Maupertuis, Ampère, Coriolis, and many more.)

Richard Olson, 'Historical reflections on feminist critiques of science: The scientific background to modern feminism', in *History of Science*, Vol 28, 1990, pp 125–47.

*Oxford Reference Encyclopedia*, Oxford University Press, Oxford, 1998.

Elizabeth Chambers Patterson, *Mary Somerville and the Cultivation of Science, 1815–1840*, Martinus Nijhoff Pubishers, The Hague, 1983.

Roger Penrose, *The Emperor's New Mind: Concerning Computers, Minds, and the Laws of Physics*, Vintage, London, 1991.

Teri Perl, *Math Equals: Biographies of Women Mathematicians and Related Activities*, Addison-Wesley, Reading, 1978.

Ivars Peterson, *Newton's Clock: Chaos in the Solar System*, W.H. Freeman & Co., New York, 1993.

John Playfair, 'Dissertation on the progress of mathematical and physical science since the revival of learning in Europe', *Encyclopaedia Britannica*, 1818.

Roy Porter and Marilyn Ogilvie (eds), *The Biographical Dictionary of Scientists*, Oxford University Press, New York, 2000.

Neil Postman, *Building a Bridge to the Eighteenth Century: How the Past can Improve Our Future*, Scribe Publications, Melbourne, 2000.

Allan F. Randall, 'Quantum superposition, necessity and the identity of indiscernibles', 1996, online preprint.

William Rankin, *Newton for Beginners*, Allen & Unwin, Sydney, 1993.

A. Rebière, *Les femmes dans la science: Conférence faite au cercle Saint-Simon le 24 février, 1894*, Librairie Nony and Co., Paris, 1894.

Robert Resnick and David Halliday, *Physics*, John Wiley & Sons, New York, 1966. (More recent editions of this excellent book are available.)

Robert Resnick, *Introduction to Special Relativity*, John Wiley & Sons, New York, 1968.

V. Frederick Rickey, 'Historical notes for the calculus classroom: History of the brachistochrone', Department of Mathematical Sciences, United States Military Academy, 1996, online.

Larry Riddle, 'Sophie Germain and Fermat's Last Theorem', Department of Mathematics, Agnes Scott College, 2009, online.

Hazel Rossotti, 'The woman that inspired Faraday', in *Chemistry World*, June 2007.

George Sebastian Rousseau, 'The consumption of meat in an age of materialism: Materialism, vitalism and some Enlightenment debates over national stereo-types', in Robert C. Leitz III and Kevin L. Cope (eds), *Imagining the Sciences: Expressions of New Knowledge in the 'Long' Eighteenth Century*, AMS Press, New York, 2004.

Jean-Jacques Rousseau, *Émile*, translated by Grace Roosevelt, from an earlier translation by Barbara Foxley, Institute for Learning Technologies, Columbia University, 1998, online.

M. Sansottera, U. Locatelli and A. Giorgilli, 'On the stability of the secular evolution of the planar Sun-Jupiter-Saturn-Uranus system', arXiv:1010.2609v1, 2010, online.

Simon Schama, *A History of Britain*, BBC television documentary series and accompanying books, 2000–2002.

Maria Susana Seguin, 'Les Femmes et les sciences de la nature', *Dix-Huitième Siècle (Femmes des Lumières)*, No 36, 2004, pp 333–43.

J. B. Shank, *The Newton Wars and the Beginning of the French Enlightenment*, University of Chicago Press, 2008.

Alan E. Shapiro, *Fits, Passions, and Paroxysms*, Cambridge University Press, Cambridge, 1993.

Quentin Smith, 'A defense of a principle of sufficient reason', *Metaphilosophy*, Vol 26, Nos 1 and 2, 1995, pp 97–106.

Mary Somerville, 'On the magnetising power of the more refrangible Solar rays', in *Philosophical Transactions of the Royal Society*, Vol 116, 1826, pp 132–9.

—, *Mechanism of the Heavens*, John Murray, London, 1831.

—, *On the Connexion of the Physical Sciences*, 7th edition, John Murray, London, 1846. It was reprinted by Arno Press, New York, 1975: references in my Notes and Sources are to this edition, and denoted by *Connexion*. (First edition by John Murray, London, 1834.)

Mary Fairfax Somerville and Martha Charters Somerville, *Personal Recollections from Early Life to Old Age of Mary Somerville, with Selections from her Correspondence*, John Murray, London, 1873. (Facsimile edition by Elibron Classics, Adamant Media Corporation, Boston, 2005.) An expanded edition of this work, with additional original writings from earlier drafts of Mary's *Recollections*, has been published as *Queen of Science* (see McMillan for details). Anne Soprani (ed.), see Châtelet, *Lettres D'Amour.* References to Soprani in my Notes and Sources refer to Soprani's introduction and endnotes to these letters.

John Stillwell, *Mathematics and its History*, Springer-Verlag, New York, 1989.

Edwin F. Taylor and John Archibald Wheeler, *Spacetime Physics*, W.H. Freeman & Co., New York, 1992.

Mary Terrall, *The Man Who Flattened the Earth: Maupertuis and the Sciences in the Enlightenment*, University of Chicago Press, Chicago, 2002. In my Notes and Sources, this reference is denoted by 'Terrall'.

—, '*Vis Viva* Revisited', *History of Science*, Vol 42, pp 189–209, 2004. In my Notes and Sources, this reference is denoted by 'Terrall (2004)'.

George B. Thomas, Jr., *Calculus and Analytic Geometry*, Addison-Wesley, Reading, 1969.

Brian Tierney and Joan W. Scott (eds), *Western Societies: A Documentary History*, Vols 1 and 2, McGraw-Hill, New York, 2002. (This valuable resource contains English versions of key historical documents.)

I. Todhunter, *A History of the Mathematical Theories of Attraction and the Figure of the Earth, from the Time of Newton to that of Laplace*, Dover Publications, New York, 1962. (This is a reprint of the original 1873 Macmillan edition.)

Scott Tremaine, 'The stability of the solar system', presentation at Princeton University's Frontiers of Astronomy School/Workshop, Bibliotheca Alexandrina, 2006, online.

René Vaillot, *Madame du Châtelet*, Albin Michel, Paris, 1978.

Voltaire, *Candide, ou l'Optimisme*, Hachette, Paris, 1972.

—, *Éléments de la Philosophie de Neuton mis à la portée de tout le monde*, Jacques Desbordes, Amsterdam, 1738.

—, *Lettres Philosophiques*, chronology and preface by René Pomeau, Garnier Flammarion, Paris, 1964.

——, *Letters Concerning the English Nations*, edited with an introduction and notes by Nicholas Cronk, Oxford University Press, Oxford, 1999.

——, *Oeuvres complètes de Voltaire* (ed. Léon Thiessé), Baudouin, Paris, 1829. These include his plays, including *Alzire*, his long essay, *Traité sur la Tolérance sur l'occasion de la Mort de Jean Calas*, and other writings mentioned in the narrative (apart from his letters, which are drawn from his *Correspondance* [ed. Besterman]).

—— correspondence: *Voltaire: Correspondance* (especially Vol 2), compiled and annotated by Theodore Besterman, Bibliothèque de la Pléaide, Éditions Gallimard, 1965. In my Notes and Sources, this reference is denoted by *Correspondance*.

Ira O. Wade, *Studies on Voltaire, with Some Unpublished Papers of Mme du Châtelet*, Princeton University Press, Princeton, 1947.

——, *The Intellectual Origins of the French Enlightenment*, Princeton University Press, Princeton, 1971.

M. Christine Walsh, *Prologue: A documentary history of Europe, 1846–1960*, Cassell, Melbourne, 1968.

Robert Weinstock, *Calculations of Variations with Applications to Physics and Engineering*, Dover Publications, New York, 1974.

Richard S. Westfall, *Never at Rest: A Biography of Isaac Newton*, Cambridge University Press, Cambridge, 1980.

Philip Wiener (ed.), *Leibniz Selections*, Scribner, New York, 1951. (This is a selection of Leibniz's original writings, translated into English.)

Mary Wollstonecraft, *Vindication of the Rights of Woman*, Miriam Kramnick (ed.), Penguin Classics, international edition, 1978.

Donald K. Yeomans, 'Comet Halley – the orbital motion', in *The Astronomical Journal*, Vol 82, No 6, 1977.

Thomas Young, *A Course of Lectures on Natural Philosophy and the Mechanical Arts*, Taylor and Walton, London, 1845. (The lectures were originally given in 1807.)

Judith Zinsser, *La Dame d'Esprit*, Penguin Books, international edition, 2006. (This book is also available in paperback as *Emilie du Châtelet: Daring Genius of the Enlightenment*.)

# ACKNOWLEDGMENTS

I would like to thank all the friends, family and readers who have encouraged me in my writing over the years, especially the incomparable Morgan – such support is a very precious gift. In terms of this book in particular, I also thank my dear friend Professor Herb Fried, of Brown University, for his critical reading and his advice on quantum processes; my 'sister in science' at Monash University, Dr Rosemary Mardling, for fascinating and helpful discussions on resonance and the stability of the solar system; and Dr Peter Hambly, of the University of Adelaide, for a meticulous eye and 'the kindness of strangers'. I also thank all the scholars, past and present, whose work has fascinated and informed me: these are explicitly acknowledged in my Notes and Sources. I am very grateful to my agents Jenny Darling and Brandi Bowles, to Donica Bettanin at JDA, and Madonna Duffy at UQP. Finally, it has been a joy to work with Timothy Bent, Executive Editor (Trade) at Oxford University Press, and I thank him and his wonderful team.

I am also extremely grateful for the financial support of the Literature Board of the Australia Council for the Arts (and Nancy Keesing, whose generosity endowed the Keesing studio in Paris), and the Arts Victoria/UNESCO City of Literature program.

# INDEX